T0155848

Essentials of Dynamics and Vibrations

John Billingsley

Essentials of Dynamics
and Vibrations

 Springer

John Billingsley
Faculty of Health, Engineering and Science
University of Southern Queensland
Toowoomba, Queensland
Australia

ISBN 978-3-319-85934-7   ISBN 978-3-319-56517-0 (eBook)
DOI 10.1007/978-3-319-56517-0

Printed on acid-free paper

This Springer imprint is published by Springer Nature
The registered company is Springer International Publishing AG
The registered company address is: Gewerbestrasse 11, 6330 Cham, Switzerland

Contents

Chapter 1
Overview

Abstract Dynamics is the study of how things move. That seems fairly obvious. But there are fundamental principles that have to be understood, despite the detail being hedged around with some pretty forbidding mathematics. We are so accustomed to Newton's laws that we cannot imagine life without them. They are easily put to the test of our experience and imagination. Yet youngsters are afflicted with pseudo-science in films that defies all reason, showing star-fighters zooming away in space just by tilting their wings or orbits 'decaying' with no air-resistance. It will be worth devoting an entire chapter to media howlers. Gyroscopes are regarded with the same mystique as zombies and vampires. The reputation of a very distinguished professor was tarnished by a belief that they could be used for space propulsion. But gyroscopes lead us unto another aspect of the theory. Newton was intrigued by linear motion, but we must also consider rotation. Just as the second law tells us to take account of linear momentum and forces, so we must additionally consider angular momentum and couples. This is where the lovers of mathematics can really come into their own!

1.1 Introduction

This is a very difficult subject, demanding mathematical ability to solve differential equations and to use and manipulate matrices. Before starting, you should visit the 'mathematical revision material' at www.essdyn.com/maths to make sure that you will survive.

You should already be familiar with the dynamics of systems with point masses. Newton's laws have told you how to deal with accelerations and forces, but now you will meet systems with a whole new twist – literally.

If you rotate a point mass, it is still a point mass. All that matters is the combination of forces that are trying to accelerate it, together with its present position and velocity.

As soon as you exchange the point for a solid body you have to start thinking in six dimensions, not three. As well as the x, y and z coordinates you have rotations of pitch, roll and yaw – to borrow some terms from aviation. As well as forces in three

© Springer International Publishing AG 2018
J. Billingsley, *Essentials of Dynamics and Vibrations*,
DOI 10.1007/978-3-319-56517-0_1

dimensions, you have to consider couples in three more. And instead of the body having a simple mass, it now has a moment of inertia, too. But there's more! That moment of inertia is actually a tensor, represented by a three-by-three matrix.

But as you work through the book, the matrix will become your friend. It will allow you to represent the mathematical relationships in a neat shorthand. It will also allow you to throw together a few lines of software to simulate any awkward system, including orbiting satellites, spinning gyroscopes and vibrating mechanisms.

Matrices also enable us to represent the transformations that describe the motion of the links of a robot.

As well as describing the motion of solid objects in three dimensions, the differential equations you meet can describe vibrations and oscillations that involve multiple modes. Once again you are thrown into the grasp of matrices to separate the individual modes and find their frequencies.

You will also have to recall the fundamentals of solving differential equations, configuring a 'complementary function' to match the initial conditions while finding a 'particular integral' to represent the response to any input function.

1.2 A Little More Detail

Dynamics is the study of how things move, not just point masses but solid objects with rotational momentum. There are fundamental principles that have to be understood, though the detail involves some challenging mathematics. At the centre are Newton's laws, which could be paraphrased as follows:

1. Unless you push it, it will just keep going at the same velocity.
2. The harder you push it, the more it will accelerate; acceleration is proportional to the push.
3. If you push it, it will push you back.

We are so accustomed to Newton's laws that we cannot imagine life without them. They are easily put to the test of our experience but films project a deluge of misinformation. They show star-fighters that zoom away in space just by tilting their wings, or orbits that 'decay' with no air-resistance. But possibly the entertainment movies have less to apologise for than the television documentaries that 'get it wrong'. It will be worth devoting an entire chapter to media howlers.

Gyroscopes are regarded with the same mystique as zombies and vampires. The reputation of a very distinguished professor was tarnished by a belief that they could be used for space propulsion. Perhaps they can, but not in any way that conforms with our reliance on Newton's laws. Though there are dangers in holding any belief too strongly, gyroscopic propulsion can be put into the same basket as perpetual motion.

But gyroscopes lead us unto another aspect of the theory. Newton was intrigued by linear motion, but we must also consider rotation. Just as the second law tells us to take account of linear momentum and forces, so we must additionally consider angular momentum and couples. This is where the lovers of mathematics can really come into their own!

Spinning objects can do remarkable things. A spinning top can stand upright. A coin can roll across the floor without falling over until it slows down, the dynamics of the front wheel of a bicycle will let us ride it 'no hands'. But irregular objects can do even more remarkable things.

If you search on the web for 'Dancing T-handle' you will see some video footage that was taken in an orbiting space station. A T-shaped bar spins 'helicopter style' for a brief while, then it suddenly flips so that the stem is pointing in the opposite direction, then repeatedly it flips back and forth again. How can we wrap our minds around that sort of motion? Well the power of the computer is there to help us.

HTML5, the 'canvas' object and JavaScript combine together to let us animate pictures on the screen of objects that obey any dynamic equations that we care to define. So at www.essdyn.com/sim/t-handle.htm we can see a simulation that explains it all.

While we can rely on simple concepts for many simulations, understanding the T-handle requires us to look deeper into the mathematics by considering the 'inertia tensor' and 'Euler's equations'. But those come much later.

Many of you readers are likely to be followers of a university course, keen to find how to pass an examination. You will want worked examples on which to try out your mathematical gymnastics. The examples in the text should keep you reasonably satisfied, provided you can resist looking at the solutions that follow them before trying to find solutions of your own. But you might miss the whole point of the book.

The mathematical methods are merely a collection of tools in your toolbox. The essence of a knowledge of dynamics is in knowing which of those tools to apply to each new situation that you might meet. It is not a matter of showing how proficient you are at matrix multiplication.

Your course leader is likely to add one or two of the 'standard titles' to your reading list. I will add a list of further reading without showing a preference for any of the titles. However I will give you a reference to the book that inspired me half a century or more ago – the fundamentals have not changed in all that time. You can download Synge and Griffith's 'Principles of Mechanics', published in 1949, from https://archive.org/details/principlesofmech031468mbp at no cost, but I do not claim that it is easy reading! It goes a lot further than these chapters.

In 1966, JL Meriam, published his book, 'Dynamics', with Wiley of New York. By 2007, writing with a partner, JL Meriam and LG Kraige 'Engineering Mechanics Vol. II Dynamics', was in its sixth edition.

Meanwhile a competing series by Russell C Hibbeler, 'Engineering Mechanics: Dynamics', has reached its thirteenth edition by 2017 and is published by Pearson. At 768 pages it is a substantial volume, with a price to match.

In my cynical view, succeeding editions have become more ornamented with illustrations until they look like a children's encyclopaedia. In term of understanding the fundamentals, a much cheaper earlier edition is likely to serve you just as well.

One virtue of such volumes is that they are packed with exercises, if you want to rehearse the same solution techniques over and over again. They can also appeal to a lecturer who is too lazy to think up original examination questions. In my opinion, however, it is much more important to gain a firm grasp of the underlying principles, rather than rehearse a set of party-trick formulae.

To continue the introduction, let us get a taste of what is to be found in the chapters you will meet.

1.3   Summary of Chapter Contents

1.3.1   Dynamics of Particles

Do not be confused! The particles here are not grains in a dust cloud, they are simply 'point masses'. In other words, they have mass but no moment of inertia and we start by considering them one at a time.

Their 'state' is just determined by their position and linear velocity; rotation does not come into the picture, unless the rotation is about some point other than their centre of mass, causing them to have a linear velocity.

The position can be expressed as a vector function of time and the velocity is given just by differentiating it. The particle's path is a locus in space and we can say a number of things about its curvature.

But though we have started with simple particles, that is just the beginning of the story. As soon as we fix a group of particles into a framework, such as at the corners of a cube, we have something with all the properties of a solid body. Now angular velocity and momentum are of great importance because that body now possesses an 'inertia tensor'. But it can all be solved by Newton's equations by relating the solid body motion to each of the particles that make it up.

On the way we have to tidy up some of the concepts arising from the move from two to three dimensions. When all the forces can be drawn on a sheet of paper, the 'triangle of forces' is enough to define linear equilibrium; when we take moments of any force to define a couple, the axis of that couple will automatically be perpendicular to the page.

But now we can have forces in any direction acting at all sorts of points on a solid body. We have to delve into the mathematics of vectors, with scalar and vector cross products and maybe meet a matrix or two.

1.3.2   Momentum and Rotary Motion

From earlier study, you should already be familiar with the idea of a moment of inertia. Newton's laws will have already told you that linear acceleration will be given by a force divided by a mass. When we make things rotate we are concerned with couples and moments of inertia. But instead of a simple number to represent the moment of inertia, when things move in three dimensions we have to consider an 'inertia tensor', a three-by-three matrix of coefficients.

Just as a force is a vector, so a couple is represented by a vector in the direction of the axis about which it twists. The integral of such a couple will be angular momentum.

But while a moving mass has linear momentum in the direction in which it is moving, a revolving body can have angular momentum that is not aligned with the axis of its angular velocity. Unless we apply a couple to that object, the angular momentum has to remain constant, which means that the angular velocity must whirl around! So how can we analyse it? That is where the inertia tensor comes in.

1.3.3 Inertia

You will have met the 'parallel axis theorem' for finding a moment of inertia when the axis of revolution is not through the centre of gravity, but now that we are considering three dimensional motion it is most straightforward if the inertia tensor is centred on the centre of mass and the motion of its centre of mass is added into the mix separately.

An object will have three mutually perpendicular principal axes of inertia about which it can spin undisturbed. For a ball these are any three perpendicular directions and it has the same moment of inertia about each. But when a more general object spins about some other axis than a major axis, strange things can happen, as you will have seen in the T-handle video.

But first in this chapter we will see the principles behind the inertia tensor, when we regard the body as an assembly of point masses. We will later see that when solid objects are combined to make a bigger object, we first calculate the inertia tensor resulting from point masses at their centres of gravity, then add on the sum of their individual inertia tensors. Of course these must all be aligned about the same set of axis directions.

So we can work out the inertia tensor of a boomerang by regarding it as a pair of sticks joined at right-angles.

1.3.4 Balancing

In Chap. 5 we will consider balancing, where your usual task is to add masses to a rotating object to minimise the bearing forces. This has a practical application when you balance the wheels of your car when buying new tyres. If you can, you make the axis of rotation a principal axis of inertia, but in other examples this may not always be possible. The inertia tensor is a useful tool to achieve this.

1.3.5 Euler's Equations

Some rather challenging mathematics concerning operators can be taken on trust, leading to a proof of Euler's equations. With these you can look at exotic behaviour such as the 'Dancing T-handle'. You can find equations to explain the behaviour of a gyroscope, something that seems a mystery to most people. You could try that boomerang example, calculating the inertia tensor and finding three axes about which it can spin without 'wobbling'.

1.3.6 Gyroscopes

Although much of the underlying theory has been established through Euler's equations, gyroscopes are considered in a little more detail. Precession is examined and it is seen that much greater analysis than that given here is needed to include nutation and the process whereby a spinning top stays erect. However to simulate that top is a much easier task.

1.3.7 Kinematics

Now we can take a rest from considering forces, couples and momentum to consider instead the geometry of a robot manipulator. Most industrial robots consist of a chain of 'revolute' axes which we can think of waist, shoulder, elbow, wrist and so on. But the position of the 'end effector', the hand that does the work, is a somewhat complicated combination of the functions that depend on all these angles.

By this stage you should be familiar with the three-by-three matrices that calculate rotations about the x, y and z axes.

For rotation about the x axis we have

$$\begin{bmatrix} 1 & 0 & 0 \\ 0 & c & -s \\ 0 & s & c \end{bmatrix}$$

while rotations about the y and z axes are given by

$$\begin{bmatrix} c & 0 & s \\ 0 & 1 & 0 \\ -s & 0 & c \end{bmatrix}$$

and

$$\begin{bmatrix} c & -s & 0 \\ s & c & 0 \\ 0 & 0 & 1 \end{bmatrix}$$

where the c and the s are the cosine and sine of the angle of the rotation.

But the transformations needed for analysing robot dynamics do not just have to represent rotations. They must also represent the translation of, for example, changing a set of axes from the shoulder to the elbow or from elbow to wrist. We could simply add the displacement to our present coordinate, but we would really like something that can be applied using the standard computer matrix multiplication routine.

So we 'fatten up' the matrix to four-by-four and add a fourth component, which is always 1, to our position vector so that from $(x, y, z)'$ it becomes $(x, y, z, 1)'$.

Now the transformation

$$\begin{bmatrix} c & 0 & s & L \\ 0 & 1 & 0 & 0 \\ -s & 0 & c & 0 \\ 0 & 0 & 0 & 1 \end{bmatrix}$$

will represent a rotation about the y axis combined with a translation L in the x direction.

1.3.8 Kinematic Chains

The next step is to string a set of four-by-four matrices together, so that we can compute the position and orientation of the end-effector in terms of the axis angles.

To avoid confusion we can separate out each individual action, such as an individual rotation or an individual translation down a limb, so that they are only combined when we multiply their matrices together.

We have to be clear about the order in which to put the matrices. If we start at the 'hand', a point with coordinates $(x, y, z, 1)$ relative to the hand would have coordinates $(x + L, y, z, 1)$ relative to an elbow, if the hand is just at a translation L in the x direction from it.

So if we think of 'travelling back from the hand', the matrices will build up from right to left. If we think of 'travelling out from the base', they will build up from left to right. So we might end up with a product of:

(waist rotate)(shoulder rotate)(translate shoulder to elbow)
(elbow rotate)(translate elbow to wrist)(wrist rotate)

and so on, in that order.

When you come to the grind of multiplying the matrices together, as long as you keep them in the right order you can pair them off in any way that you wish, starting from the right or left or even in the middle.

1.3.9 Inverse Kinematics

This is one of the more challenging tasks. If you are designing a robot controller, you will have the problem of 'inverse kinematics'. You will know the desired position and attitude of the end-effector, but to control your robot you have to

calculate a set of axis values to correspond to them. The Jacobian can give ways
of closing in on a solution without having to solve the algebra.

1.3.10 Vibration

Up to this point you will have had to brush up your matrix algebra. But now the
mathematics comes in thick and fast.

Hang up a pendulum or attach a mass to the end of a spring and you have 'simple
harmonic motion'. This chapter is all about vibration.

When there is a force or couple that accelerates something towards some central
position, things vibrate.

- If there is no 'damping' to soak up energy, they will vibrate forever.
- If the force or couple is proportional to the displacement, then the movement
  will be a sinusoidal (sine or cosine) function of time.
- If there is some damping, with an additional force or couple that is proportional
  to velocity, the sine-wave will 'decay' exponentially.

As an engineer, you will probably wish to add some damping to limit some
unwanted vibration or to move the 'resonant' frequency away from that of a disturbance such as a spinning motor.

As a musician, you would wish to shape the vibration to something that sounds
pleasant! So to deal with a vibrating system you should:

1. Look for the 'variables' that describe what is happening.
2. Find some equations for their rates-of-change in terms of all such variables and
   any inputs.
3. Eliminate all the variables but one, to get a differential equation.
4. Solve this equation to analyse what will happen.

Now there are two ways to go about step 3. You can mess about with simultaneous equations and algebra, or you can use the power of matrices to help you.

5. Write the 'state equations' in matrix form.
6. Look for 'eigenvalues' and 'eigenvectors', since the matrix product that gives
   the rate-of-change of an eigenvector is just the eigenvalue times the vector itself.

This gives an exponential solution that is easy to recognise. But if there is
damping the exponentials can be complex! They can then represent the product of
a sine-wave with a decaying exponential.

And so you have entered the territory of an electronic engineer, where complex
exponentials are to be expected all the time. You will have to brush up on your
complex arithmetic, in addition to considering particular integrals and complementary functions.

But as soon as you add a second mass, the complications move up a notch.

1.3.11 Modes

Spotting a pair of state variables can enable you to analyse a single-degree-of-freedom system that involves just a single frequency, perhaps a complex one. An example of such a system is a simple pendulum, swinging from side to side. But that pendulum has another degree of freedom where it swings towards us or away, or any combination of the two, swinging diagonally or in circles or ellipses. It has two frequencies, although in this case they are the same. The two orthogonal directions of swing are its 'modes', sinewaves with different phases which are added together to define the position.

In general there can be numerous modes. Consider a rectangular steel plate, suspended vertically by springs at the corners of its top edge.

An object moving in three dimensions has six degrees of freedom – three of position and three more of rotation. So there will be six ways in which the plate can vibrate or swing. These are:

1. Up and down bouncing on the springs.
2. Rotary twisting on the springs about an axis perpendicular to its plane.
3. Sideways swinging in a vertical plane.
4. Rotation about a vertical axis.
5. A combination of tilting with swinging towards us and away, where the mass tilts in the same direction as the swing.
6. Another combination where the tilt opposes the swing.

These can all happen at the same time!

Each of the first four can be analysed individually, using the techniques you will have used for single degree of freedom systems. But the last two require something more special.

An example that is easier to visualise concerns two trolleys of equal mass on a frictionless track. They are connected by three equal springs. One of these joins the trolleys, the other two extend from each end, connecting each of the trolleys to a fixed 'anchor point'.

There are two modes. In one the trolleys move in unison, in the other they bounce together and apart. You can see a simulation showing the modes as follows:

1. At www.essdyn.com/sim/modes1.htm you can see one of the modes – after you click on 'run the model'.
2. At www.essdyn.com/sim/modes2.htm you can see the other.
3. At www.essdyn.com/sim/modes3.htm you can see what happens when both modes are excited at the same time.

So how can we unscramble them?

If at this stage we are not interested in any damping, we can look for the two second-order differential equations that give their accelerations. Since each of the accelerations involves both of the positions, the equations can be expressed in matrix form. Now we look for eigenvalues and eigenvectors and we have a solution for the modes. The frequencies will be the square roots of the negative eigenvalues.

But as soon as you add some damping, the mathematics will move up to another level.

First we must identify the four 'state variables', two positions and two velocities of the trolleys.

Then we must write an equation for the rate-of-change of each of these.

We must express these as a first-order matrix equation in the vector holding these four variables.

Then we must work out the 'characteristic equation' in lambda (fourth order in this case) that gives us the eigenvalues when we solve for its set of roots.

If there is no damping, the 'fourth order' equation simplifies to a quadratic in the square of lambda.

If there is damping that is not 'over critical', the roots will fall into complex conjugate pairs, representing a sine-wave multiplied by an exponential.

If all this mathematics worries you, do some urgent revision! You can find some help at www.essdyn.com/maths.

1.3.12   Rocket Science!

Around you is a dynamic world. From a tender age, you have been exposed to the cinema film director's concepts of dynamics – and so many of them are wrong!

When a 'starfighter' banks, it is shown zooming away in a curve. But Newton's laws tell you that you have to fire a rocket perpendicular to the path if you want to change its line. The star-ship 'Enterprise' is shown with the unfortunate property that if its engines fail, its orbit will decay rapidly. But it is only the satellites in an orbit so low that air drag takes effect that need have any worry.

If the orbit starts to decay, the satellite can be boosted into a higher orbit. A UK documentary showed the jets propelling the satellite straight up. Is this really the correct answer?

But even more hokum is talked about Black Holes, as though they are giant vacuum cleaners gobbling suns and linking to 'wormholes'. The concept of 'gravity gradient' can change your view.

Some literary critics hail Jules Verne as a scientific genius. But his books are full of mathematical and engineering howlers.

Jules Verne has the excuse of having written it long ago, but the same cannot be said for the perpetrators of the film 'Gravity' that seems to break every kinetic law in the book.

1.4 Summary

So as well as the ability to string mathematics and software together to solve problems of forces, couples, resonances and mechanisms, you will be able to see the world clearly in terms of the equations that make things move, stay still or perhaps fall down.

Technical points will be illustrated by software so simple that it will run inside your Internet browser and make dynamic images on your screen. The software is completely open, residing in a web page. 'View source' will open everything up so that you can edit it to make simulations of your own. However these demonstrations are intended to clarify the dynamics to you, you will not be required to write any software.

Chapter 2
Particle Kinematics and Dynamics

Abstract We enter the world of three dimensions gently, considering particles which just have mass, without the moments of inertia of a solid body. We see that vector functions of position can be differentiated to give vector velocities and accelerations. We define frames of reference and see how the motion of a particle in a rotating frame can be represented in 'world axes'. Generalised coordinates are mentioned, though a more mathematical treatment is to be found in the second appendix.

2.1 Laws and Axes

We are so used to Newton's laws that we take them for granted – ignoring any thoughts of relativity. They are easy to put to the test of our experience and imagination. In an appendix we can find Lagrangean and Hamiltonian methods that achieve their objectives without assuming Newton's laws, but for now let us stick to the clear and simple.

The first law is really contained in the second law. Put simply the laws are as follows:

1. Unless you push it, it will just keep going. Or in Newton's own words, 'Every body perseveres in its state of being at rest or of moving uniformly straight forward, except insofar as it is compelled to change its state by forces impressed.'
2. The harder you push it, the more it will accelerate; acceleration is proportional to the push. Or as Newton put it, 'A change in motion is proportional to the motive force impressed and takes place along the straight line in which that force is impressed.'
3. If you push it, it will push you back. In Newton's words, 'To any action there is always an opposite and equal reaction; in other words, the actions of two bodies upon each other are always equal and always opposite in direction.'

Actually these are translations of Newton's words. For the first law he stated, 'Every thing doth naturally persevere in yt state in wch it is unless it bee

© Springer International Publishing AG 2018   13
J. Billingsley, *Essentials of Dynamics and Vibrations*,
DOI 10.1007/978-3-319-56517-0_2

interrupted by some externall cause, hence a body once moved will always keepe ye same celerity, quantity & determination of its motion.'

Of course we will want to wrap these up into mathematical expressions that we can manipulate to calculate the answer to some problem. If we turn the second law around, it becomes

$$\mathbf{F} = \mathrm{ma}$$

where $\mathbf{F}$ is the force, $\mathbf{a}$ is the acceleration and for now we will regard the mass m as constant.

But why have $\mathbf{F}$ and $\mathbf{a}$ been written in bold type? Because they are vectors. They have both magnitude and a direction that must be aligned to match.

In other texts, you might find that vectors are instead denoted by being underlined, or perhaps have an arrow printed above them. But the problems of representing a vector do not stop there.

In most of this book, we will represent vectors in a Cartesian form, in terms of their components in three orthogonal (mutually perpendicular) directions. We can represent these directions by three unit vectors $\mathbf{i}$, $\mathbf{j}$ and $\mathbf{k}$ and represent a vector as something like $2\mathbf{i} + 3\mathbf{j} + 4\mathbf{k}$.

But in two dimensions we could represent the vector just by a distance and a heading, such as 'a hundred metres north-east'.

For manipulating the mathematics, it is usually convenient to leave $\mathbf{i}$, $\mathbf{j}$ and $\mathbf{k}$ as understood, and simply write the vector in the form

$$\begin{bmatrix} 2 \\ 3 \\ 4 \end{bmatrix}$$

Why do we use this awkward column vector? It is so that using the conventional way for multiplying vectors and matrices we can write a product as 'matrix times vector' rather than 'vector times matrix'. But we can add a dash or a superscript T to mean 'transpose' and type it more neatly on a line as

$$(2, 3, 4)' \text{ or } (2, 3, 4)^{\mathrm{T}}.$$

Just to confuse matters further, we usually label the axes x, y and z and think of the vector as $(x, y, z)'$.

If we are going to use three orthogonal unit vectors, we find that there are two possibilities, 'right handed' or 'left handed'. The convention is to use the right-handed set.

Many books illustrate this with a picture of a hand. Although I am very well aware which is my right hand and which is my left, I can never remember which of my fingers and thumb should be the x, y or z axis. Instead I think of a sheet of graph paper.

Fig. 2.1 Graph paper

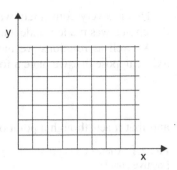

Now if we were to define the z axis to be downwards we would have a left-handed set of axes. It is the right-handed set that we need, with z upwards.

Fig. 2.2 Graph with added
z axis

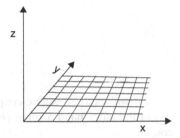

That settles matters nicely for positions and displacements, but we will also have rotations and couples to consider.

We have become used to the fact that a positive rotation in the x-y plane is anticlockwise.

Fig. 2.3 Positive rotation
about z

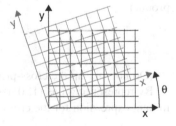

Now a rotation in three dimension must also be considered as a vector. The direction of the vector will represent the axis about which the rotation takes place. So this anticlockwise rotation will be represented by a positive vector in the z direction. That is, it is anticlockwise when we look inwards along the z axis.

But if we look outwards along the z-axis, viewing the x-y plane from below, the rotation will appear to be clockwise! If we combine the rotation theta with a positive movement in the z-direction, we will advance just like a right-handed corkscrew. Not surprisingly, this combination of rotation and translation is called a 'screw'!

This is a very convenient way of thinking of it. Do you believe that historically the choice was made accidentally, or did the corkscrew affect the definition?

As well as rotations we have to think in terms of couples, twists around an axis. Suppose that we have a force in the y direction, which we could write as

$$(0, F_y, 0)'$$

and that it acts though a point on the x-axis that we can write as $(x, 0, 0)'$.

Fig. 2.4 A force giving a positive couple

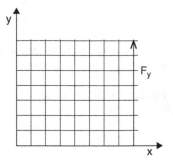

This will exert a positive couple about the origin – anticlockwise looking down.

On the other hand a force $(F_x, 0, 0)$ acting at $(0, y, 0)$ would exert a negative couple.

We can add them to get a resulting couple

$$(x\, F_y - y\, F_x)\, \mathbf{k}$$

In general the couple produced by force $\mathbf{F}$ through position $\mathbf{x}$ will be the vector product

$$\mathbf{x} \times \mathbf{F}$$

(Do not mix up the cross-product symbol '$\times$' with the vector $\mathbf{x}$!)

By representing $\mathbf{i}$ as $(1, 0, 0)'$, $\mathbf{j}$ as $(0, 1, 0)'$ and $\mathbf{k}$ as $(0, 0, 1)'$, using the determinant expression for the cross product, you can show that

$$\mathbf{i} \times \mathbf{j} = \mathbf{k}$$

$$\mathbf{j} \times \mathbf{k} = \mathbf{i}$$

$$\mathbf{k} \times \mathbf{i} = \mathbf{j}$$

and also that

$$\mathbf{i} \times \mathbf{k} = -\mathbf{j}$$

2.2   Particle Kinematics

Later on we will consider solid bodies, but for now we are looking at particles.

What is a particle?

It is something we can consider to be described just by its position, without any need to consider which way it is oriented or twisted. In the simulation that you can find at www.essdyn.com/sim/orbitxyz.htm a satellite is represented by a simple point, moving through space, though it is portrayed as a ball.

Below the picture of the orbit, you will see two windows of code. With the program stopped, you can edit the initial position and velocity in the upper box.

If you make the values of z and vz zero, you will see a 'flat' orbit.

In the lower box, you can also change the code that calculates and updates the position. It just represents the condition that the acceleration of the planet is inversely proportional to the square of its distance from the 'sun', acting in the same direction as the vector from the planet to the sun.

The planet's position is represented by a three-dimensional vector $(x, y, z)'$.

As you will see when you click on the link, x, y and z vary with time. They could be computed in a simulation, as in this case, or for a different example they could be specified explicitly as functions of time, such as

$$\text{position} = \mathbf{x}(t) = (t, t^2, t^3)'$$

The bold x signifies that we are dealing with a vector, not just the x component.

The particle will have a velocity that we can get by differentiating its position vector term by term:

$$\text{velocity} = d\mathbf{x}(t)/dt = (1, 2t, 3t^2)'$$

We can differentiate this again to get the acceleration:

$$\text{acceleration} = d^2\mathbf{x}/dt^2 = (0, 2, 6t)'$$

(Often you will see the time derivative denoted by putting a dot over the x for the first derivative $\dot{\mathbf{x}}$, two dots for the second derivative $\ddot{\mathbf{x}}$ and so on.)

The particle will travel along a path or 'locus'. If there is no acceleration, this path is a straight line – just as Isaac Newton says.

If the acceleration is directed along the path, it is still a straight line but the particle will get faster or slower.

Only if we direct a thrust at an angle to the line will we get it to curve. (This might come as a surprise to the makers of space-flight movies such as Star Wars or Star Trek!)

If we take the line of the thrust and the local line of the path, they will define a plane. This is called the 'osculating plane' and is a sort of super-tangent to the curve. The centre of curvature will lie in that plane.

Now in the orbit example, the satellite will keep moving around in a plane. But in general the acceleration could vary in direction and the plane could twist around the curve as the particle goes along it.

2.3   Vector Derivatives in Rotating Systems

We are entering the thorny ground of mathematics. Now would be a good time to look at Appendix 1 ('Mathematicians and operators') to get up to speed on partial derivatives and sets of rotating axes.

You have to get your mind around the concept of partial and total derivatives.

Imagine that you are sitting on a children's roundabout and that you have chalked some axes on it. You start at the origin and with your vector displacement **r** you move along the x-axis at 10 cm/s. That speed of $(0.1, 0, 0)'$ is the partial derivative which we can write as

$$\frac{\partial \mathbf{r}}{\partial t}$$

We have not yet said whether the roundabout is turning or stationary.

But now let us suppose that it is turning at a rate ω and that you are at vector distance **r** from the centre. The rotation of the roundabout will then give you an additional velocity which is the cross product of vector ω with the vector **r**. Your total velocity relative to world axes is

$$\frac{d\mathbf{r}}{dt} = \frac{\partial \mathbf{r}}{\partial t} + \omega \times \mathbf{r}$$

So if the position vector is **r** then the velocity is not just the rate-of-change of **r** as measured relative to the rotating axes, we must also add a velocity term due to **r** 'swinging around the axis' to get this total velocity.

The mathematicians think in terms of an 'operator'

$$\frac{d}{dt} = \frac{\partial}{\partial t} + \omega \times$$

When we use this to differentiate the velocity, we see that the acceleration is

$$\frac{d\mathbf{v}}{dt} = \frac{\partial \mathbf{v}}{\partial t} + \omega \times \mathbf{v}$$

but we have used the operator twice to differentiate the position, so

$$\frac{d^2\mathbf{r}}{dt^2} = \left(\frac{\partial}{\partial t} + \omega \times\right)\left(\frac{\partial}{\partial t} + \omega \times\right)\mathbf{r}$$

which we can expand to get

$$\frac{d^2\mathbf{r}}{dt^2} = \left(\frac{\partial}{\partial t} + \omega \times\right)\left(\frac{\partial \mathbf{r}}{\partial t} + \omega \times \mathbf{r}\right)$$

$$= \frac{\partial}{\partial t}\left(\frac{\partial \mathbf{r}}{\partial t} + \omega \times \mathbf{r}\right) + \omega \times \left(\frac{\partial \mathbf{r}}{\partial t} + \omega \times \mathbf{r}\right)$$

$$= \frac{\partial^2 \mathbf{r}}{\partial t} + \frac{\partial \omega}{\partial t} \times \mathbf{r} + \omega \times \frac{\partial \mathbf{r}}{\partial t} + \omega \times \left(\frac{\partial \mathbf{r}}{\partial t} + \omega \times \mathbf{r}\right)$$

$$= \frac{\partial^2 \mathbf{r}}{\partial t^2} + \frac{\partial \omega}{\partial t} \times \mathbf{r} + \omega \times \frac{\partial \mathbf{r}}{\partial t} + \omega \times \frac{\partial \mathbf{r}}{\partial t} + \omega \times (\omega \times \mathbf{r})$$

$$= \frac{\partial^2 \mathbf{r}}{\partial t^2} + \frac{\partial \omega}{\partial t} \times \mathbf{r} + 2\omega \times \frac{\partial \mathbf{r}}{\partial t} + \omega \times (\omega \times \mathbf{r})$$

- The first term is the acceleration with respect to the axes.
- If the rotation of the axes is accelerating, the second term tells us that our velocity will also have an acceleration term proportional to the angular acceleration.
- The third term is the Coriolis force.
- The final term is the 'centripetal acceleration' (commonly thought of as resulting in centrifugal force) directed towards the axis and perpendicular to it, the product of the resolved component of $\mathbf{r}$ perpendicular to the axis and the square of the angular velocity.

You will mostly find the examples in Appendix 1, but this seems to be a useful one to include in the text.

2.4   Roundabout Poser

You are riding on a roundabout in a children's playground, standing on the edge of the platform, facing towards the centre. You are 2 m from the centre.

The roundabout is rotating anticlockwise at 2 rad/s.

You swing your right leg in and out at 1 m/s. Your lower leg has mass 2 kg.

1. Are you likely to kick your left leg on the in-swing or the out-swing?
2. How much force is required to keep the swing on a straight line from the centre?
3. How much centripetal acceleration do you experience?
4. Your friend is sitting on the platform, 1.5 m from the centre, not holding on. If the coefficient of friction is 0.5, will they slide off?

Solution

1. You will be travelling to your right (looking inwards).

    Your foot has greater velocity when extended behind you than forwards (assuming it is on the radius) so swinging it inwards will seem to push it to your right.

    You will kick your left ankle on the out-swing.

    Try it!
2. The Coriolis force is twice the mass times omega times $dr/dt = 2 \times 2 \times 2 \times 1 = 8$ N.
3. The centripetal acceleration is 2 m radius times omega squared = 8 m/s^2.
4. Your friend at 1.5 m from the centre will have acceleration $r\,\omega^2 = 1.5 \times 2^2 = 6$ m/s$^2 = 0.6$ g.

The friction force will only be 0.5 g times the mass, so the friend will indeed slide off!

2.5 Motion of a Particle in a Moving Coordinate System

So far we have assumed that although the axes are rotating, their origin is fixed in space. But in general, everything can be happening at once! We might want our axes to be defined in terms of a spaceship that is moving, accelerating and spinning.

If the position of the origin of the coordinates is **R**, then we simply add **R** to the position **r**, and the first and second derivatives of **R** to the velocity and acceleration.

But we must be careful. Of course we can easily add vectors in a conceptual sense, but the calculations of velocities and accelerations will give the answers in terms of axes that are at this instant aligned with the moving coordinate system. To add the derivatives of **R** we will also expect **R** to be measured in directions aligned with that coordinate system. But in many cases we will need our answers to be expressed relative to an 'absolute frame of reference', perhaps a set of axes fixed in the world. (And for now we ignore the fact that the world is itself rotating and moving!)

As soon as we want to change our frame of reference, we have to consider transformations. If you have read the vector notes in Appendix 1 carefully, you will already be familiar with ways to transform sets of axes.

2.6   Generalized Coordinates and Degrees of Freedom

There are many ways to analyse a system. We can apply Newton's Laws to obtain equations for accelerations, from which we can build the 'state equations' that are at the heart of a simulation. We can instead look at conservation of energy to get a more general hold on variables that might be hard to express in Cartesian terms.

Lagrange had an approach that involved minimising the 'action', the difference between the kinetic and potential energy. Hamilton instead used the sum of potential and kinetic energies to get something that was a bit more intuitive.

Lagrange also had a method of dealing with constraints. This has been termed 'Lagrange's undetermined multipliers'. You will find a wealth of information by Googling, but the aspects that are really relevant to these topics can be found here or in Appendixes 1 and 2.

Suppose that a particle can move in three dimensions, but is constrained to slide along a wire. It obeys the usual expression for kinetic energy and its potential energy in this case is just gravity times its height.

Without the presence of the wire, we would simply solve equations to find that the particle accelerates vertically downwards. The wire makes all the difference!

If it is in the shape of a vertical circle, the particle behaves like a pendulum, swinging to and fro at the bottom. If it is a horizontal straight line, the particle does not accelerate at all.

Now a free particle would have three degrees of freedom. The wire reduces it to just one. There are actually two constraints.

A function like

$$f_1(x, y, z) = 0$$

represents a surface, even if nonlinear, so we need a second constraint

$$f_2(x, y, z) = 0$$

so that the wire can be defined as the intersection of these surfaces.

Lagrange adds these equations into the mixture that must be solved for the motion.

But instead of dealing with the three variables x, y and z, we could just use some measure of the distance travelled by the particle along the wire. We would then only have one variable to deal with.

This is the principle of 'generalised coordinates'.

You can download Newton's 'Principia' from Project Gutenberg at http://www.gutenberg.org/ebooks/28233.

Chapter 3
Linear and Angular Momentum

Abstract The two important conservation laws are conservation of energy and conservation of momentum. From the concept of linear momentum and forces, impacts can be considered. 'Newton's Cradle' and the 'Galilean Cannon' are each illustrated with a simple JavaScript simulation. But we must also deal with angular momentum and couples. It is shown that a displacement and a rotation can be combined into a rotation about another axis, or into a 'screw'. In the same way, a couple and a force can either be combined into a force with a different line of action, or more generally into a 'wrench'. Finally linear and angular momentum are combined with an impulse in finding the 'sweet spot' of a cricket bat.

3.1 Linear Momentum

The momentum of an object is the product of its mass with its vector velocity. The momentum **p** is defined by

$$\mathbf{p} = m\mathbf{v}$$

Newton's second law states that for the momentum **p** of an object to change, some force must be applied to it. We can express the rate of change in mathematical terms as

$$d\mathbf{p}/dt = \mathbf{F}$$

This is just a restatement of Newton's law

$$\mathbf{F} = m\mathbf{a}$$

Now this force does not have to come from outside the entire system that we are considering. It can be a force between two elements in contact. Whether you consider that the 'equal and opposite' properties of such forces are a consequence of Newton's third law, or whether you use the techniques of the 'free body dia-gram', you have the result for two interacting particles that

$$\mathrm{d}\mathbf{p}_1/\mathrm{d}t + \mathrm{d}\mathbf{p}_2/\mathrm{d}t = 0$$

unless some external force is applied to them.

Indeed, if our system has a whole lot of particles within it, whether they form a machine with pistons, cranks and casings or whether they are just a wobbly jelly, we can write

$$\sum \left(\frac{\mathrm{d}\mathbf{p}_i}{\mathrm{d}t} \right) = \mathbf{F}_{ext}$$

or

$$\frac{\mathrm{d}}{\mathrm{d}t} \sum (\mathbf{p}_i) = \mathbf{F}_{ext}$$

This just means that if we add up the momentum of every bit of mass in our system, then the rate-of-change of the total is whatever force we are exerting from outside it.

An obvious consequence is that if our system has some unbalanced vibrating element within it, the changes in momentum will result in vibration of the case, or at least cause dynamic forces in its mountings. The only complete remedy is to add opposing dynamic elements so that the total momentum remains constant. In other words, we must carry out dynamic balancing.

Elements within a spaceship can vibrate and spin, but the momentum of the system as a whole will only be changed by external forces such as gravitational attraction.

But in that case, how does a spaceship propel itself?

Burning gases are hurled backwards out of the rocket tubes. Considered as a whole, the momentum of spaceship including the expelled gases will still be con-stant, but when we drop the momentum of these gases from the system, the result is that the spaceship with its remaining fuel is given forward momentum.

3.2 Impacts with Point Masses

Two point masses can interact 'gently' – as when two charged objects come close together, or they can collide with an impact. In theory, the forces needed for a step change in momentum will be infinite for an infinitesimal time. In practice, there is elastic deformation, possibly with waves of compression running through the objects – but the result is just the same as a step change.

So how can we model an impact?

At www.essdyn.com/sim/newton.htm there is a simple simulation of **Newton's Cradle**. For some time these have been sold as executive desktop toys.

Balls in a row are suspended so that they can swing against each other. If a pair of balls swing together against the left-hand end of the row, a pair of balls will be sprung away from the right hand end. Click on a ball to see it happen.

The effect is just the result of a sequence of impacts.

Another simulation is a simplified **billiards game** at www.essdyn.com/sim/billgame.htm.

Here the balls can move in two dimensions, and we have to look at the closing velocity not only in terms of its magnitude, but also in terms of its direction. So for the 'bounce' we add the appropriate vector velocity to each ball.

It is obvious what to do about an impact when the masses are equal, but what do we do when they are different?

The total momentum must remain unchanged, so the change made to each of the masses involved must alter the momentum by an equal and opposite amount − the change in velocity must be inversely proportional to each mass.

We can illustrate this with yet another simulation.

At www.essdyn.com/sim/cannon.htm there is a demonstration called a **Galilean Cannon** that you can try yourself.

Hold a ping-pong ball on top of a golf ball, drop them together and see the ping-pong ball bounce high.

There is a YouTube example that you can find by a web search for 'Galilean Cannon'.

Run the simulation after you have tried the real experiment!

What is the theoretical maximum height it can reach, if I can select a variety of masses for the balls? First calculations suggest nine times the starting height, but try 40 for the ratio and be surprised. Of course the real balls could not be dropped accurately enough to keep bouncing perfectly in line!

3.3   Angular Momentum

Just as we were able to sum the mass-times-velocities and hence the momentum of all the particles and get a total linear momentum, so we can take their moments to get their angular momentum

$$\sum (\mathbf{r}_i \times \mathbf{p}_i)$$

The problem lies in the vector $\mathbf{r}_i$. Where is it measured from? A single point mass rushing along a line will have an angular momentum about any point that is not on the line.

So we will measure the angular momentum from the centre of gravity of the object that concerns us, adding another term if we want to know the momentum about some other point.

We now have laws that look very much like Newton's laws, except that instead of linear momentum and forces they now concern angular momentum and couples.

If we call the total angular momentum $\mathbf{L}$, we have

$$\mathbf{L} = \sum m_i(\mathbf{r}_i \times \mathbf{v}_i)$$

and

$$\frac{\mathrm{d}\mathbf{L}}{\mathrm{d}t} = \tau$$

where vector tau is the applied torque.

3.4 Forces, Couples, Translations and Rotations

First let us consider everything to lie in a plane.

We can first consider one 'transformation' that simply rotates an object,

Fig. 3.1 Rotation

then another that moves it from one place to another, or yet another that does both.

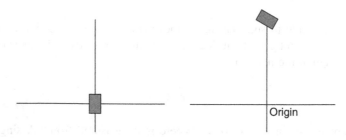

Fig. 3.2 Rotation plus translation

But in most cases this combination can be turned into a single rotation about some other point.

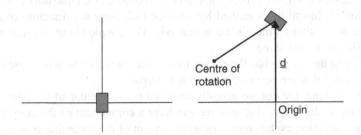

Fig. 3.3 Rotation about 'another point'

Obviously this is not possible if there is translation without any rotation – the centre of rotation would be at infinity.

So what has this to do with forces and couples?

First let us define a couple. It is the result of two equal opposing forces that do not act along the same line. When added together there is no net resulting force, but there is a twist. What might seem odd is that this twist is not associated with any particular location, it acts anywhere in the plane.

So in all the cases illustrated below, the couple is the same.

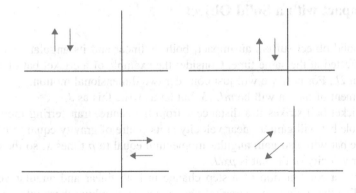

Fig. 3.4 Couples

If the forces are half the size but twice as far apart, the result is the same.

Now just as we were able to combine a displacement with a rotation to get a rotation about a different point, so we can combine a force with a couple to get a force acting through a different point.

We will see the importance of this when we consider gyroscopes.

3.5   Three Dimensions

When we advance to three dimensions, all this changes. A couple now has to have a 'direction'. It has to be described by a vector indicating the direction of an axis to operate about, though there is no actual axis. The couple about any axis parallel to that axis will be the same.

Once again the couple has the power to relocate forces, or at least those components of a force that are perpendicular to the couple.

In the x-y plane, the couple actually operated perpendicular to the plane, while the force lay in the plane. But now we can have a component of the couple that is in the same direction as the force – or a component of the force that is in the direction of the couple, whichever way we want to look at it.

The result is that we can 'dispose' of the perpendicular component of the couple by moving the force, but we are left with a couple about the line of the force – this combination is called a **wrench**. It is really more like the action of a screwdriver!

A similar problem occurs with our transformations.

When the axis of rotation is perpendicular to a displacement, we can remove the displacement by making the rotation happen about a different centre. But in three dimensions the rotation axis and the displacement can have a parallel component. We are left with an axis of rotation and a displacement along that axis. This is called a **screw**.

3.6   Impact with a Solid Object

When a solid object suffers an impact, both its linear and its angular momentum will be affected at the same time. Consider the example of a cricket bat of mass m and length $2L$. For now we will just consider two dimensional motion.

Its moment of inertia will be $mL^2/3$, but let us write this as J.

If a cricket ball strikes it a distance x from the centre, transferring momentum p, the whole bat will gain a linear velocity at its centre of gravity equal to p/m.

But the bat will also gain angular momentum equal to p times x, so the change in angular velocity of the bat is px/J.

We have a combination of a step change in both linear and angular velocity, and recalling the topic in paragraph 4 above, we can combine these into a rotation about some other point. This point will be at distance $-(p/m)/(px/J)$ from the centre of gravity.

With a bit of simplification we can express this as $-(J/m)/x$. If the impact is below the centre of gravity the rotation centre will be above it.

Now sometimes the moment of inertia is defined in terms of the **radius of gyration** R to be $J = mR^2$.

In this case R would be $L/\sqrt{3}$.

But that means that we can define the centre of rotation more simply as $-R^2/x$.

It is called the **centre of percussion**.

Exercise 3.1

A batsman holds the (lightweight) handle of the bat a distance $L/2$ above the body of the bat, which is of length $2L$.

He hits the ball for a six and the handle does not sting his hand. At what location on the bat did the ball make contact?

Solution

If the ball strikes a distance x below the centre of the bat, the 'centre of percussion' will be at a distance R^2/x above it. But we know that $R^2 = L^2/3$ and that the bat is held $3L/2$ above the centre.

So since

$$(L^2/3)/x = 3L/2 : \text{where the bat is held}$$

$$x = (L^2/3)/(3L/2) = 2L/9$$

the 'sweet spot' where the impulse of the ball does not impart any impulse to the hand of the batsman will be just a little below the centre of the bat.

Chapter 4
Inertia

Abstract When we make things rotate we are concerned with **couples** and **moments of inertia**. Just as a force is a vector, so a couple is represented by a vector in the direction of the axis about which it rotates. The integral of such a couple will be **angular momentum**. But while a moving mass has linear momentum in the direction in which it is moving, a revolving body can have momentum that is not aligned with the axis of its angular velocity. That is where the **inertia tensor** comes in. An object will have three mutually perpendicular principal axes of inertia about which it can spin undisturbed. For a ball these are **any** three perpendicular directions, having the same moment of inertia about each. But when a more general object spins about some other axis than a major axis, strange things can happen. But first in this chapter we will see the principles behind the inertia tensor, when we regard the body as an assembly of point masses. We later see that when solid objects are combined to make a bigger object, we first calculate the inertia tensor resulting from point masses at their centres of gravity, then add on the sum of their individual inertia tensors. Of course these must all be aligned about the same set of axis directions. So we can work out the inertia tensor of a boomerang by regarding it as a pair of sticks joined at right-angles.

4.1 Introduction

In previous studies you will have considered motion in two dimensions with rotation in a plane. In three dimensions, the moment of inertia **J** becomes an **inertia tensor**. It is not as alarming as it sounds! But first we should perhaps revise some of the rules for the simple case.

4.2 Why Do We Need a Tensor?

We started by considering the momentum of individual point masses. Now we are going to lock several of them together to form a solid body. This has a centre of mass which in turn can have a velocity **v** while the whole body can rotate about the centre of mass with angular velocity **omega**.

© Springer International Publishing AG 2018
J. Billingsley, *Essentials of Dynamics and Vibrations*,
DOI 10.1007/978-3-319-56517-0_4

Let us start simple, without that velocity **v** and with just rotation.

Let us consider two point masses, each of mass m, at $(-r, 0, 0)$ and $(+r, 0, 0)$ on the x axis. What is their moment of inertia? You will probably answer $2mr^2$. But that is the two-dimensional answer where there is just one moment of inertia, about an axis that is perpendicular to the paper.

But if this object spins about the x-axis, its moment of inertia is zero! The masses stay on the x axis and do not move.

About the y and z axes, the moment of inertia is indeed $2mr^2$, but it is clear that we cannot just represent inertia by a single number. We are about to find that we cannot even use a vector, we have to use a tensor, in this case a three-by-three matrix. Let us call it **J**.

We want to use it to calculate angular momentum. When the object rotates with vector velocity ω we hope to calculate the angular momentum as

$$\mathbf{J}\omega$$

Now in the Chap. 3 we saw that moments of the linear momentum can be taken to get the angular momentum.

So now let us consider another two-mass object, this time with the first mass at $(-r, -r, 0)$ and mass 2 at $(r, r, 0)$. Once again its centre of mass will be at the origin. We will take both masses as unity.

Fig. 4.1 Two point masses

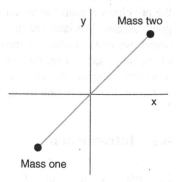

We will spin it about the x axis with angular velocity $(\omega, 0, 0)'$.

As we look down from the z direction we can see that instantaneously the first mass is moving away from us with velocity $(0, 0, -r\omega)'$ while mass 2 is moving towards us with velocity $(0, 0, r\omega)'$. So to compute the angular momentum we add $(-r, -r, 0) \times (0, 0, -r\omega) + (r, r, 0) \times (0, 0, r\omega)$.

The result is $(r^2\omega, -r^2\omega, 0) + (r^2\omega, -r^2\omega, 0) = (2r^2\omega, -2r^2\omega, 0)$. The momentum is not in the same direction as the angular velocity.

As the masses whirl around the x axis, so the momentum will also whirl around. This changing momentum will have to be provided by a whirling couple, provided by the bearing forces, so the system is decidedly not balanced!

One way to compute that couple is by struggling with matrices. The other is to consider the 'centrifugal forces'. Mass 1 will apply a negative force in the y direction of $(0, -r\omega^2, 0)$ while mass 2 will apply $(0, r\omega^2, 0)$. When we add up the moments of these two forces about the origin we get $(0, 0, 2r\omega^2)$.

This is the couple exerted by the masses, so the bearings will have to apply $(0, 0, -2r\omega^2)$. And remember that this is the couple at time $t = 0$, before the masses have moved out of their starting position. As the masses rotate, the forces and the couple will rotate, too. But as the answers get littered with sines and cosines of ωt we will start to realise that the matrix approach does not have so many drawbacks, after all.

At the same time we realise that $\mathbf{J}$ cannot be a scalar or even a vector, since in this case

$$\mathbf{J}(\omega, 0, 0)' = (2r^2\omega, -2r^2\omega, 0)'$$

and it takes a matrix or a tensor to change the direction of a vector.

4.3   Angular Momentum and the Inertia Tensor

To find out how to calculate the inertia tensor for a collection of point masses, we are going to have to grit our teeth and get to grips with some mathematics. But the result will be a formula with which we can 'crank the handle' to perform the calculation.

Let us consider a general solid object which we regard as a collection of any number of point masses, i.e. particles. It is rotating about its centre of mass with constant angular velocity. The ith particle is located at $\mathbf{r}_i$ and its linear velocity is given by

$$\mathbf{v}_i = \omega \times \mathbf{r}_i$$

So the angular momentum (about the centre of mass) of this individual particle is given by

$$m_i \mathbf{r}_i \times (\omega \times \mathbf{r}_i)$$

Now according to the mathematicians, this *vector triple product* can be expanded as

$$m_i(\omega(\mathbf{r}_i \cdot \mathbf{r}_i) - \mathbf{r}_i(\omega \cdot \mathbf{r}_i))$$

The first term is quite straightforward. It is just what we expect from our 'in-the-plane' experience of omega times the moment of inertia mr^2, now aligned with omega, the axis of rotation.

But the second term suggests that if $\mathbf{r}_i$ and omega are not orthogonal, the angular momentum can have a component in the direction of $\mathbf{r}_i$.

So for the x component we get

$$m_i(\omega_x \mathbf{r}_i^2 - (\mathbf{r}_i)_x(\omega \cdot \mathbf{r}_i))$$
$$= m_i(\omega_x \mathbf{r}_i^2 - (\mathbf{r}_i)_x \{\omega_x(\mathbf{r}_i)_x + \omega_y(\mathbf{r}_i)_y + \omega_z(\mathbf{r}_i)_z\})$$
$$= m_i(\omega_x \{(\mathbf{r}_i)_y^2 + (\mathbf{r}_i)_z^2\} - (\mathbf{r}_i)_x \{\omega_y(\mathbf{r}_i)_y + \omega_z(\mathbf{r}_i)_z\})$$

and when we do the summation to get the total, leaving off the i subscripts, we have

$$\mathbf{L}_x = \omega_x \sum m(\mathbf{r}_y^2 + \mathbf{r}_z^2) - \omega_y \sum m\mathbf{r}_x\mathbf{r}_y - \omega_z \sum m\mathbf{r}_x\mathbf{r}_z$$

But this is just the x component.

For the whole vector $\mathbf{L}$ we must take the product of the vector ω with a matrix – or as the mathematicians would call it, a tensor.

Let us apply an intuitive test to this. When things just lie in a plane, so that the z components of all the $\mathbf{r}$'s are zero, we look at just the z component of the momentum.

$$\mathbf{L}_z = -\omega_x \sum m\mathbf{r}_x\mathbf{r}_z - \omega_y \sum m\mathbf{r}_y\mathbf{r}_z + \omega_z \sum m(\mathbf{r}_x^2 + \mathbf{r}_y^2)$$

Because omega has only a z component, the first two terms are zero. So we are left with an expression with which we are already familiar.

For the whole inertia tensor $\mathbf{J}$ we have

$$\mathbf{J} = \begin{bmatrix} \sum m(\mathbf{r}_y^2 + \mathbf{r}_z^2) & -\sum m\mathbf{r}_x\mathbf{r}_y & -\sum m\mathbf{r}_x\mathbf{r}_z \\ -\sum m\mathbf{r}_x\mathbf{r}_y & \sum m(\mathbf{r}_x^2 + \mathbf{r}_z^2) & -\sum m\mathbf{r}_y\mathbf{r}_z \\ -\sum m\mathbf{r}_x\mathbf{r}_z & -\sum m\mathbf{r}_y\mathbf{r}_z & \sum m(\mathbf{r}_x^2 + \mathbf{r}_y^2) \end{bmatrix}$$

so that the angular momentum in the three dimensional case is

$$\mathbf{L} = \mathbf{J}\omega$$

If you have not followed every detail of the mathematics, it does not really matter. You will just have to take the formula for $\mathbf{J}$ on trust.

Exercise 4.1
Consider that simple rotor with just two unit point masses.
   One is at $(1, 1, 0)$ and the other is at $(-1, -1, 0)$.
   The rotor is simulated at www.essdyn.com/sim/unbalanced.htm.

1. Work out the values in the inertia tensor.
   Use it to answer the following:
   A rotational velocity ω is applied about the x-axis.
2. What is the angular momentum?

3. What is the axis of the momentum?
4. Does this make sense? (Compare the result with a simple calculation without matrices.)
5. What would the inertia tensor be if the masses were at $(-1, 1, 0)$ and $(1, -1, 0)$?

Solution
1. *Work out the values in the inertia tensor.*

$$\mathbf{J} = \begin{bmatrix} \sum m(\mathbf{r}_y^2 + \mathbf{r}_z^2) & -\sum m\mathbf{r}_x\mathbf{r}_y & -\sum m\mathbf{r}_x\mathbf{r}_z \\ -\sum m\mathbf{r}_x\mathbf{r}_y & \sum m(\mathbf{r}_x^2 + \mathbf{r}_z^2) & -\sum m\mathbf{r}_y\mathbf{r}_z \\ -\sum m\mathbf{r}_x\mathbf{r}_z & -\sum m\mathbf{r}_y\mathbf{r}_z & \sum m(\mathbf{r}_x^2 + \mathbf{r}_y^2) \end{bmatrix}$$

Here you can see the simple 1 or -1 numbers being put into the expression for **J** to give you a matrix, in this case

$$\begin{bmatrix} 2 & -2 & 0 \\ -2 & 2 & 0 \\ 0 & 0 & 4 \end{bmatrix}$$

2. *What is the angular momentum?*
For the second part you just have to post-multiply the matrix by the vector angular velocity

$$\mathbf{J} \, (\omega, 0, 0)'$$

(i.e. column vector to the right of the matrix) to get the momentum (column vector again)

$$(2\omega, -2\omega, 0)'$$

3. *What is the axis of the momentum?*
This is clearly not in the same direction as the angular velocity, but is in the direction

$$(1, -1, 0)'$$

which is at $-45°$ to the axis.
4. *Does this make sense? (Compare the result with a simple calculation without matrices.)*

Now if you see the white ball heading towards you and the red ball heading away, their instantaneous velocities would just as well be given by a rotation about that axis at −45° – it looks sensible.

5. *What would the inertia tensor be if the masses were at (−1, 1, 0) and (1, −1, 0)?*

For part 5 you move the balls and you should get an answer

$$
\begin{bmatrix}
2 & 2 & 0 \\
2 & 2 & 0 \\
0 & 0 & 4
\end{bmatrix}
$$

4.4   Angular Momentum and Couples

The rate-of-change of the angular momentum of a body is the couple that is applied to it. In a form that looks a bit like Newton's second law $\mathbf{F} = d\mathbf{p}/dt$, this can be written as

$$
\mathbf{C} = \frac{d\mathbf{L}}{dt}
$$

It enables us to calculate the bearing forces from the couple that we must apply to keep that angular velocity constant. So we have

$$
\mathbf{C} = \frac{d(\mathbf{J}\omega)}{dt}
$$

$$
= \mathbf{J}\frac{d\omega}{dt} + \frac{d\mathbf{J}}{dt}\omega
$$

$$
= \frac{d\mathbf{J}}{dt}\omega
$$

since

$$
\omega = \begin{bmatrix} \omega \\ 0 \\ 0 \end{bmatrix}
$$

where omega is constant, so its derivative is zero.

So we must express $\mathbf{J}$ as a function of time.

Now if at time $t = 0$ the masses in our example were at $(1, 1, 0)'$ and $(-1, -1, 0)'$ and we are rotating them about the x axis with angular velocity ω, then at time t they will be at

$$\begin{bmatrix} 1 \\ \cos \omega t \\ \sin \omega t \end{bmatrix}$$

and

$$\begin{bmatrix} -1 \\ -\cos \omega t \\ -\sin \omega t \end{bmatrix}$$

We can calculate $\mathbf{J}$, differentiate it and then multiply it by $(\omega, 0, 0)'$ to get the couple.

If we locate the bearings at $(1, 0, 0)'$ and $(-1, 0, 0)'$ we can work out the forces they need to apply.

4.5 Energy

We can calculate the kinetic energy of an object composed of point masses by adding up their individual energies:

$$\sum \frac{1}{2} m |\mathbf{v}_i|^2$$

$$= \sum \frac{1}{2} m |\omega \times \mathbf{r}_i|^2$$

As you might expect, the same inertia tensor is involved when we unscramble the expression for the energy. But now we have to multiply the tensor with the vector omega twice to get the scalar energy. The result is a *quadratic form*

$$\frac{1}{2} \omega' \mathbf{J} \omega$$

What happens if we apply a transformation to $\mathbf{J}$, as happens when we rotate our system about the x axis?

Applying a transformation to ω would be simple. But suppose we now consider an angular velocity $\mathbf{T}^{-1}\omega$ (which is the same as $\mathbf{T}'\omega$, since from our knowledge of rotation matrices we know that $\mathbf{T}'$ is the same as $\mathbf{T}^{-1}$).

The transpose of $\mathbf{T}'\omega$ is $\omega'\mathbf{T}$ so the energy now appears as

$$\frac{1}{2} \omega' \mathbf{T} \mathbf{J} \mathbf{T}' \omega$$

So why did we consider the rotation $\mathbf{T}'\omega$ instead of $\mathbf{T}\omega$?

The energy is a scalar, so that transformations cannot affect it. We could either consider transforming $\mathbf{J}$, while ω stays the same, or if we look from the aspect of those new axes, $\mathbf{J}$ stays the same but the angular velocity now appears as $\mathbf{T}'\omega$.

Instead of just multiplying $\mathbf{J}$ by $\mathbf{T}$ to transform it, we must transform it with

$$\mathbf{TJT}'$$

Exercise 4.2

(Exercise 4.1 continued)

6. Calculate the inertia tensor as a function of time, where the point masses are now at $(1, \cos \omega t, \sin \omega t)$ and $(-1, -\cos \omega t, -\sin \omega t)$.
7. Calculate the couple required to keep the body rotating about the x axis.
8. If the bearings are at $(1, 0, 0)$ and $(-1, 0, 0)$ what are the forces on them?
9. Calculate the forces on the bearings by considering the two masses individually, in terms of centrifugal force. Do your answers match?
10. Using the inertia tensor, what is the kinetic energy of the system?
11. What is the kinetic energy when you just consider the two masses? Is the answer the same?

Solution

6. *Calculate the inertia tensor as a function of time, where the point masses are now at $(1, \cos \omega t, \sin \omega t)$ and $(-1, -\cos \omega t, -\sin \omega t)$.*

   For part 6 we consider the balls as they make their way around their circular paths. It is very similar to the earlier solution, but now you have some sines and cosines in your expression.

   To keep it neat we will write c for $\cos(\omega t)$ and s for $\sin(\omega t)$:

   $$\begin{bmatrix} 2 & -2c & -2s \\ -2c & 2+2s^2 & -2cs \\ -2s & -2cs & 2+2c^2 \end{bmatrix}$$

7. *Calculate the couple required to keep the body rotating about the x axis.*

   Now in part 7 it gets just a bit trickier, because we need to differentiate J omega.

   But ω is constant, so we just need dJ/dt omega.

   What is more, we only need the first column of dJ/dt,

since it is multiplied by $(\omega, 0, 0)'$, so the first column of dJ/dt is

$$(0, 2\omega \, s, -2\omega \, c)'$$

and for the product we get an extra omega

$$(0, \ 2\omega^2 \, s, -2\omega^2 \, c)'$$

8. *If the bearings are at (1, 0, 0) and (−1, 0, 0) what are the forces on them?*
   For part 8, we consider the forces that could give us such a couple –
   say by taking moments about each bearing point a and b.

   The bearings are distance 2 apart, so the couple is 2 times force at a,
   or 2 times force at b.

   We can assume that the forces along the axis in the x direction are
   zero – they will give no moment.

   We can find the forces without any reference to matrices:
   We find that the bearing a at $(1, 0, 0)'$ must apply a force

   $$(0, -\omega^2 \, c, -\omega^2 \, s)'$$

   to counter the 'centrifugal force' of the mass, while the other one applies

   $$(0, \omega^2 \, c, \omega^2 \, s)'$$

   That fits the answer of part 7 – and so solves part 9.

9. *Calculate the forces on the bearings by considering the two masses
   individually. Do your answers match?*
   Now for part 9 we consider the effect of the centrifugal force exerted by
   each of the masses, acting in the same direction as their current (y, z)
   position.

   And provided we realise that the force on the bearing is equal and
   opposite to the force applied by the bearing, we find the same answer.

10. *Using the inertia tensor, what is the kinetic energy of the system?*
    Part 10 involves a simple bit of matrix multiplication, but

    $$(\omega, 0, 0) \, J \, (\omega, 0, 0)'/2$$

    only involves the top left coefficient of J, giving the scalar value ω^2.

11. *What is the kinetic energy when you just consider the two masses? Is
    the answer the same?*
    And in part 11 it is no surprise that each of the two masses, travelling at
    speed 1 times ω, will have energy $\omega^2/2$, which when added together
    gives the same answer of omega squared.

So for your 'model answer' you would need just a few words of explanation, far less than I have given here.

The lesson I am trying to put across is that matrices give a useful 'crank the handle' method of solving dynamics problems, while the answers have to be the same as the result of performing the calculations piece by piece.

4.6 Comments and Clarification of the Exercises

When you calculate the angular momentum you get a vector, where the direction of the vector represents an axis about which the momentum acts. Just as a rate of change of linear momentum defines a force, so rate of change of angular momentum defined a couple. Once again an axis is involved.

If the object spins about an axis about which it is dynamically balanced, the rotation and the momentum are aligned, so no couple is required to keep it going. If the momentum is not aligned with the axis, then it will be 'whirling around' with a non-zero rate of change, so the bearings have to exert a couple.

Now in the case above, the positions of the 'rotated' masses are easy to calculate. But suppose a mass starts at a more 'awkward' position at $t = 0$, such as $(1, 1, 1)'$, where will it be at time t, rotated by an angle ωt?

If you look at the maths revision notes at www.essdyn.com/maths and find the section on 'orientation' you will see that to rotate the point about an axis we multiply its coordinate vector by a matrix. For example a rotation about the x axis will move the point that was at $(1, 1, 1)'$ to

$$\begin{bmatrix} 1 & 0 & 0 \\ 0 & c & -s \\ 0 & s & c \end{bmatrix} \begin{bmatrix} 1 \\ 1 \\ 1 \end{bmatrix}$$

where $c = \cos(\omega t)$ and $s = \sin(\omega t)$.

The simulation at www.essdyn.com/sim/4Dmatdeg.htm will show you matrix multiplications in action. In this case, the x axis will be the roll axis.

Now there are many ways to get to an answer.

1. You can work out the dynamics of the individual masses,
2. You can transform the masses to work out the inertia matrix $\mathbf{J}$ at time t, then multiply by the velocity vector to get the angular momentum, then differentiate that to get the couple
3. You can work out the inertia matrix at $t = 0$, as in the earlier exercise, then transform it with $\mathbf{T'JT}$, where $\mathbf{T}$ is the matrix just above, then work out the

angular momentum by multiplying it by the vector $(\omega, 0, 0)'$ and differentiate to get the couple.

4. You can calculate the momentum at $t = 0$, with the nice, simple **J**. This will give you a vector. Now you know that this vector will 'whirl' around the x axis. So you can simply multiply this vector by the matrix above to find its value at time t. Then you can differentiate that to get the couple. Much easier!

4.7   Some Questions for Practice

Question 4.1
The dynamics of a communications satellite are dominated by those of its solar panel. This is a rectangular plate 4 m by 2 m. Its mass can be denoted by M.

(a) If the z axis is taken perpendicular to the center of the plate, while the x axis is in the center of the long direction, what is the satellite's inertia tensor?

(b) Show that the inertia can be represented by four equal point-masses on two of the axes. What are their locations?

Solution
(a) For a stick of length L about its centre, the moment of inertia is $mL^2/12$. The plate can be imagined to consist of a stack of such sticks, so about the long bisector the moment of inertia will be $2^2/12$ while about the short bisector it will be $4^2/12$.

Now from first principles, the first of these would involve $\sum x^2 \delta m$ while the second would involve $\sum y^2 \delta m$.

If we add these together we get $\sum (x^2 + y^2)\delta m$ which is $\sum r^2 \delta m$, just what we need for the moment of inertia about the perpendicular axis z.

So the inertia tensor is found to be

$$\begin{bmatrix} 1 & 0 & 0 \\ 0 & 4 & 0 \\ 0 & 0 & 5 \end{bmatrix} \text{ times } m/3$$

(b) We would get the same result from four masses of $m/4$ at

$$(0, \pm 2/\sqrt{6}, 0) \text{ and } (0, \pm 4/\sqrt{6}, 0)$$

Question 4.2

A different solid body consists of three masses located at $(0, 2, 2)'$, $(-1, -1, -1)'$ and $(1, -1, -1)$.

(a) What is its inertia tensor?
(b) This new body spins about each of the axes in turn with angular velocity ω. What is the momentum vector in each case?
(c) Is the body dynamically balanced about any of these axes? (Is the angular momentum vector aligned with the axis of rotation?)
(d) The body must be able to spin freely about the z-axis. Two more unit masses are to be added at $(x, y, 1)'$ and $(-x, -y, -1)'$ to restore dynamic balance. What are the values of x and y?
(Hint: You might consider the inertia tensor of this pair of masses.)

Solution

(a) First add the coordinates (times their respective masses) to be sure of the location of the centre of mass.
Cranking the numbers into the point mass formula gives us

$$J = \begin{bmatrix} 8+2+2 & 0-1+1 & 0-1+1 \\ 0-1+1 & 4+2+2 & -4-1-1 \\ 0-1+1 & -4-1-1 & 4+2+2 \end{bmatrix}$$

i.e.

$$J = \begin{bmatrix} 12 & 0 & 0 \\ 0 & 8 & -6 \\ 0 & -6 & 8 \end{bmatrix}$$

(b) The body is only balanced about the x axis

$$J(\omega, 0, 0)' = (12\omega, 0, 0)'$$

$$J(0, \omega, 0)' = (0, 8\omega, -6\omega)' \quad - \text{ not aligned with the velocity}$$

$$J(0, 0, \omega)' = (0, -6\omega, 8\omega)' \quad - \text{ also not aligned.}$$

(c) As mentioned, the x axis.

(d) For masses at $(x, y, 1)'$ and $(-x, -y, -1)'$, the inertia tensor would be

$$\begin{bmatrix} 2(y^2 + 1) & -2xy & -2x \\ -2xy & 2(x^2 + 1) & -2y \\ -2x & -2y & 2(x^2 + y^2) \end{bmatrix}$$

We need to cancel the -6 term while avoiding introducing any new off-axis terms. So

$$-2y = +6$$
$$-2xy = 0$$
$$-2x = 0$$

It is clear that $x = 0$, $y = -3$ will give the answer we need.

4.8   Combining Solid Objects

Suppose that we have several solid objects clamped together to make a bigger object. Assume for instance that the communication satellite above actually has a heavy rod, perpendicular to one corner. How do we work out the inertia tensor for the combination?

Each component will have a mass and an inertia tensor. There will be a new centre of gravity, found just by considering the components as point masses at their individual centres of gravity. So we can derive the following rules:

1. Calculate the new centre of gravity, regarding the components as point masses.
2. If necessary, transform all the individual inertia tensors to align them in the directions of the same set of axes.
3. Using a new set of axes through the new centre of mass, calculate new coordinates for those point masses relative to that centre of mass.
4. Calculate the inertia tensor for the assembly of point masses (which are at the centres of mass of the elements).
5. Add to this tensor the sum of the individual (transformed) tensors of the elements. That gives the final answer.

Example 4.1
Let us consider that boomerang. Let the sticks be of unit mass and of length 4 (to make the numbers easy). They lie along the x and y axes, joining at the origin.

Calculate the inertia tensor and find the principal axes. Try it yourself before you read the solution.

Solution

The moment of inertia of a stick about its centre is $mL^2/12$, which for each of our sticks gives 4/3 (since the sticks are of unit mass and length 4).

The stick along the x axis will have zero moment about x, so its inertia tensor is

$$\begin{bmatrix} 0 & 0 & 0 \\ 0 & 4/3 & 0 \\ 0 & 0 & 4/3 \end{bmatrix}$$

while the tensor for the y-axis stick will be

$$\begin{bmatrix} 4/3 & 0 & 0 \\ 0 & 0 & 0 \\ 0 & 0 & 4/3 \end{bmatrix}$$

But now we have to consider the tensor for the point masses located at the centres of mass of the sticks. We must start by representing these centres as relative to the centre of mass of the whole boomerang.

Relative to the origin, the centres are at $(2, 0, 0)'$ and $(0, 2, 0)'$. So the boomerang's centre of mass is at $(1, 1, 0)$. Taking this as our new origin, the stick centres are at $(1, -1, 0)$ and $(-1, 1, 0)$. These are the coordinates we use in the equation for J, getting

$$\begin{bmatrix} 2 & 2 & 0 \\ 2 & 2 & 0 \\ 0 & 0 & 4 \end{bmatrix}$$

So when we add them all together we get

$$\begin{bmatrix} 10/3 & 2 & 0 \\ 0 & 10/3 & 0 \\ 0 & 0 & 20/3 \end{bmatrix}$$

Now we must find the principal axes, which means looking for eigenvalues and eigenvectors, usually requiring us to solve a third order equation in lambda. But that is not always the case.

If we multiply the matrix by $(1, 1, 0)'$ we can see that the result $(16/3, 16/3, 0)'$ is the same vector multiplied by 16/3.

Similarly multiplying by $(1, -1, 0)'$ gives $(4/3, -4/3, 0)'$ so the corresponding eigenvalue is 4/3, the principal moment about that vector.

Multiplying by $(0, 0, 1)'$ gives $(0, 0, 20/3)'$ and we have the third principal moment of value 20/3.

So the boomerang can spin freely about either of two diagonal axes in the xy plane, or about a perpendicular z axis, spinning as a flat projectile when aimed at a kangaroo.

Example 4.2

A solid hemisphere of mass m and radius r is placed with its axis of symmetry along the positive x-axis and with its flat face on the plane $x = 0$.

Its centre of gravity is at $x = 3r/8$.

A point mass m is placed at $x = -3r/8$ to balance the system.
What is the inertia tensor?
(The moment of inertia of a sphere about its centre is $0.4\ mr^2$.)

Solution

First you need to calculate the inertia tensor for a pair of point masses m at $(3r/8, 0, 0)$ and $(-3r/8, 0, 0)$, using the formula for **J**.

Then you need to add the inertia tensor of a hemisphere about its own centre of mass.

So how do you calculate the tensor for a hemisphere? You can grind it out the hard way, or look for an easy way.

You know the tensor for a sphere.

You can consider the sphere as an assembly of two hemispheres, where its tensor will be the sum of the tensor for the pair of point masses at their centres of mass, plus two 'self tensors' for their individual rotations about their own centres of mass.

Instead of working out the detailed algebra, we can manipulate these tensors as symbols. We just have to give them 'names'.

So we will call the tensors **S** for the sphere, **P** for the pair of point masses and **H** for each individual hemisphere's 'self tensors'. Now we already know the tensor for a sphere and we can work out the tensor **P** for the point masses. The sphere's tensor will be the sum of **P** plus the two hemisphere self-tensors, i.e.

$$\mathbf{S} = \mathbf{P} + 2\,\mathbf{H} - \text{so we can work out } \mathbf{H}$$

$$\mathbf{H} = (\mathbf{S} - \mathbf{P})/2$$

What we are looking for is $\mathbf{P} + \mathbf{H}$, the point mass pair plus only one hemisphere's 'self inertia'.

Let us call this answer **A**,

$$\mathbf{A} = \mathbf{P} + \mathbf{H}$$
$$\mathbf{A} = \mathbf{P} + (\mathbf{S} - \mathbf{P})/2$$
$$\mathbf{A} = \mathbf{P}/2 + \mathbf{S}/2.$$

Nice and intuitive! It's half that of the sphere's tensor plus half of the point-mass tensor.

For **P** we have

$$\begin{bmatrix} 0 & 0 & 0 \\ 0 & \frac{9}{64} & 0 \\ 0 & 0 & \frac{9}{64} \end{bmatrix} \text{ times } 2\, mr^2$$

(since there are 2 point masses each of mass m) 9/64 is around 0.14 at a guess.

For **S**/2 we have half of 2/5 $2m\, r^2$ for each axis

$$\begin{bmatrix} 0.4 & 0 & 0 \\ 0 & 0.4 & 0 \\ 0 & 0 & 0.4 \end{bmatrix} mr^2.$$

(since the whole sphere would be $2m$).

So for the total we get

$$\begin{bmatrix} 0.4 & 0 & 0 \\ 0 & 0.54 & 0 \\ 0 & 0 & 0.54 \end{bmatrix} mr^2$$

Exercise 4.3

The dynamics of a different communications satellite are dominated by those of its solar panel and a rod antenna. This panel is a rectangular plate 4 m by 4 m. Its mass can be denoted by M.

(a) If the z axis is taken perpendicular to the centre of the plate, while the plate is square to the x and y axes, what is the plate's inertia tensor?

(b) If the rod also has mass M and extends from the $(2, 2, 0)'$ corner 2 m in each of the positive and negative z directions, where is its centre of gravity and what is its inertia tensor about that centre of gravity?

(c) Where is the centre of mass of the whole satellite?

(d) What is the inertia tensor of the pair of point masses representing plate and rod about that centre of gravity?

(e) What is the inertia tensor of the whole satellite about axes through its centre of gravity? (Axes in the same directions as the original axes?)

(f) What are the principal axes of inertia and what is the inertia tensor about those axes?

Solution

(a) *If the z axis is taken perpendicular to the center of the plate, while the x axis is in the center of the long direction, what is the panel's inertia tensor?*
You should certainly recall that the moment of inertia of a rod about its end is $1/3\ ML^2$ and about its centre is $1/12\ ML^2$ – or at least you should know where to look it up!

By imagining a stack of rods side by side, making a square, the moment of inertia about an axis across their centres will also be $1/12\ ML^2$ where M is now the total mass.

Since the plate extends 2 m either side of the x-axis, that is in the y direction, we have moment about the x axis of $4/3\ M$.

For the moment about the y-axis we also have $4/3\ M$, since the width in the x direction is 4.

Now the moment of inertia about the z axis will be the sum of these. Why? Because to calculate the moment by integration for each of these axes we are integrating elements of mass times x or y squared, whereas for z we must integrate those elements times r^2, where r is the distance from the axis, and $r^2 = x^2 + y^2$.

So for the inertia about the z axis we have the sum of these two, $8/3\ M$
The inertia tensor will be

$$\begin{bmatrix} 4 & 0 & 0 \\ 0 & 4 & 0 \\ 0 & 0 & 8 \end{bmatrix} \text{ multiplied by } M/3$$

(b) *If the rod also has mass M and extends from the $(2, 2, 0)'$ corner 2 m in both positive and negative z directions, where is its centre of gravity and what is its inertia tensor about that centre of gravity?*
The centre of gravity of the rod will be at $(2, 2, 0)$, at its centre.

As in the case of the plate, we have $1/12\,M$ length^2 about each of the rod's x and y axes and zero about its length along the z-axis. So we get

$$\begin{bmatrix} 4 & 0 & 0 \\ 0 & 4 & 0 \\ 0 & 0 & 0 \end{bmatrix} \text{multiplied by } M/3$$

(c) *Where is the centre of gravity of the whole satellite?*

With the plate centre of mass M at $(0, 0, 0)'$ and the rod's centre of mass M at $(2, 2, 0)$ we have a total mass $2M$ at a centre of gravity half way between, at $(1, 1, 0)'$.

(d) *What is the inertia tensor of the pair of point masses about that centre of gravity?*

We now have to express the centres of mass of the two elements in terms of the new CoG.

Relative to that new centre, the mass coordinates are $(1, 1, 0)$ for the rod and $(-1, -1, 0)$ for the plate.

So using the formula we get M times

$$\begin{bmatrix} 1^2 + 1^2 & -1*1*1 - 1*1*1 & -1*0 \\ -1*1*1 - 1*1*1 & 1^2 + 1^2 & -1*0 \\ -1*0 & -1*0 & 1^2 + 1^2 + 1^2 + 1^2 \end{bmatrix}$$

i.e.

$$\begin{bmatrix} 2 & -2 & 0 \\ -2 & 2 & 0 \\ 0 & 0 & 4 \end{bmatrix} M$$

or for later use

$$\begin{bmatrix} 6 & -6 & 0 \\ -6 & 6 & 0 \\ 0 & 0 & 12 \end{bmatrix} \text{multiplied by } M/3$$

(e) *What is the inertia tensor about axes through its centre of gravity, in the directions of the original axes?*

Now we have them all with a common denominator, so we have

$$\begin{bmatrix} 4+4+6 & 0-6 & 0 \\ 0-6 & 4+4+6 & 0 \\ 0 & 0 & 8+0+12 \end{bmatrix} \text{times } M/3$$

which gives

$$\begin{bmatrix} 14 & -6 & 0 \\ -6 & 14 & 0 \\ 0 & 0 & 20 \end{bmatrix} \text{multiplied by } M/3$$

(f) *What are the principal axes of inertia and the inertia tensor about those axes?*

This is where the arithmetic could get nasty!

To get the eigenvalues, we take the determinant of

$$\begin{vmatrix} 14-\lambda & -6 & 0 \\ -6 & 14-\lambda & -6 \\ 0 & 0 & 20-\lambda \end{vmatrix}$$

which expands very neatly by the bottom row to give

$$(20-\lambda)((14-\lambda)^2 - 36) = 0$$

From this it is clear that one root is $\lambda = 20$, while the others are given by

$$\lambda^2 - 28\lambda + 160 = 0$$

giving another root of $\lambda = 20$ and one of $\lambda = 8$.

So the principal moments of inertia are 20 $M/3$, 20 $M/3$ and 8 $M/3$.

The three principal axes are found as the eigenvectors corresponding to the three eigenvalues.

At a glance it can be seen that $(0, 0, 1)'$ is an eigenvector, corresponding to one of the 20 $M/3$ moments.

You can also see that $(1, -1, 0)'$ is another eigenvector, corresponding to the other 20 $M/3$ moment and that $(1, 1, 0)'$ corresponds to 8 $M/3$ – just multiply the vector by the tensor and the result is obvious.

Exercise 4.4

1. What is the inertia tensor of a 1 metre square flat panel of mass M?
2. If four such panels are connected together to make a 2-m square, what is its inertia tensor?
3. Compare this result with that for a 2-m square panel of mass $4M$.
4. One of the panels breaks off. What is the inertia tensor of the assembly of the remaining three? (Try it two ways, first by putting the three panels together, then by adding a 'negative' panel to the original assembly of four.)

5. What can you say about the axes about which this three-panel assembly should be able to spin smoothly?
6. A mass M is added to replace the missing panel at the location that was that panel's centre of mass. What is the new inertia tensor? Is it the same as the original one?

Solution

1. *MI of 1 m panel*

$$\begin{bmatrix} 1 & 0 & 0 \\ 0 & 1 & 0 \\ 0 & 0 & 2 \end{bmatrix} \text{times } M/12$$

2. *MI of four such panels.*
   Let the panels be arranged as

$$\begin{array}{cc} C & A \\ D & B \end{array}$$

Their centres will be at

$A=(0.5, 0.5, 0)'$, $B=(0.5, -0.5, 0)'$, $C=(-0.5, 0.5, 0)'$ and
$D=(-0.5, -0.5, 0)'$

First we must regard them as four point masses.
    When you enter numbers into the formula for J, you find that contributions from point masses at A and D each give negative off-diagonal terms with

$$\begin{bmatrix} 1 & -1 & 0 \\ -1 & 1 & 0 \\ 0 & 0 & 2 \end{bmatrix} \text{times } M/4$$

(As a test we try the matrix on $(1, 1, 0)'$ and find that the answer is zero. Since this is the diagonal through the two masses, about which we know the *MI* is zero, this is OK.)
    The contributions from point masses at B and C give positive terms. When all four are added together these off-axis terms cancel, just giving

$$\begin{bmatrix} 1 & 0 & 0 \\ 0 & 1 & 0 \\ 0 & 0 & 2 \end{bmatrix} \text{times } M$$

Now to get the total tensor we first add up four times the tensor for a single panel about its centre, being

$$\begin{bmatrix} 1 & 0 & 0 \\ 0 & 1 & 0 \\ 0 & 0 & 2 \end{bmatrix} \text{times } M/3$$

and when we add the point-mass tensor we get

$$\begin{bmatrix} 4 & 0 & 0 \\ 0 & 4 & 0 \\ 0 & 0 & 8 \end{bmatrix} \text{times } M/3$$

3. *A 2-m panel of mass 4M will give us*

$$\begin{bmatrix} 4 & 0 & 0 \\ 0 & 4 & 0 \\ 0 & 0 & 8 \end{bmatrix} \times 4M/12$$

so of course it is the same.

4. *One of the panels breaks off …*

Method 4.1a

If it is the $A = (0.5, 0.5, 0)'$ panel that breaks off, the new centre of mass will be at

$$(-1/6, -1/6, 0)'$$

So relative to this new CoG, we have three small-panel centres

$$(2/3, -1/3, 0)', \ (-1/3, 2/3, 0)' \text{ and } (-1/3, -1/3, 0)'$$

So the point mass tensor will be

$$\begin{bmatrix} 1+4+1 & 2+2-1 & 0+0+0 \\ 2+2-1 & 4+1+1 & 0+0+0 \\ 0+0+0 & 0+0+0 & 5+5+2 \end{bmatrix} \text{times } M/9$$

i.e.

$$\begin{bmatrix} 6 & 3 & 0 \\ 3 & 6 & 0 \\ 0 & 0 & 12 \end{bmatrix} \text{times } M/9$$

i.e.

$$\begin{bmatrix} 8 & 4 & 0 \\ 4 & 8 & 0 \\ 0 & 0 & 16 \end{bmatrix} \text{times } M/12$$

Now we must add three panels-worth of 'self inertia'

$$\begin{bmatrix} 1 & 0 & 0 \\ 0 & 1 & 0 \\ 0 & 0 & 2 \end{bmatrix} \text{times } M/12 \times 3$$

which when added will give

$$\begin{bmatrix} 8+3 & 4 & 0 \\ 4 & 8+3 & 0 \\ 0 & 0 & 16+6 \end{bmatrix} \text{times } M/12$$

i.e.

$$\begin{bmatrix} 11 & 4 & 0 \\ 4 & 11 & 0 \\ 0 & 0 & 22 \end{bmatrix} \text{times } M/12$$

With 11's in it, this might look too nasty to be right! Lots of chances for a slip. But we can check with the second method.

Method 4.1b

This is a crafty attempt to avoid so much arithmetic. It is not really necessary and if you do not understand it you should keep clear of it.

We know from part 1 that the self-tensor for a 1 metre plate is

$$\begin{bmatrix} 1 & 0 & 0 \\ 0 & 1 & 0 \\ 0 & 0 & 2 \end{bmatrix} \text{times } M/12$$

and that for the 2 metre plate it is

$$\begin{bmatrix} 4 & 0 & 0 \\ 0 & 4 & 0 \\ 0 & 0 & 8 \end{bmatrix} \text{times } 4M/12$$

i.e.

$$\begin{bmatrix} 16 & 0 & 0 \\ 0 & 16 & 0 \\ 0 & 0 & 32 \end{bmatrix} \text{times } M/12$$

Now instead of calculating the sum of three plates, let us consider a 2 metre plate, with centre at $(0, 0, 0)'$ to which is added a **negative** 1 metre plate with centre at $(0.5, 0.5, 0)'$

The new centre of mass will be at $4M*(0, 0, 0)' + -1M*(0.5, 0.5, 0)'$ divided by $3M$, which is the new total mass.

So again we find the centre to be at $(-1/6, -1/6, 0)'$.

But now for the point-mass contribution we must put $4M$ at $(1/6, 1/6, 0)'$ and $-M$ at $(4/6, 4/6, 0)'$ – yes, their centres are both moved in the same direction to be positive of the new centre of mass origin.

So our calculation of the point-mass J formula gives us

$$\begin{bmatrix} 4*1 - 1*16 & -1*4 + 1*16 & 0 \\ -1*4 + 1*16 & 4*1 - 1*16 & 0 \\ 0 & 0 & 4*2 - 1*32 \end{bmatrix} \text{times } M/36$$

giving

$$\begin{bmatrix} -12 & 12 & 0 \\ 12 & -12 & 0 \\ 0 & 0 & -24 \end{bmatrix} \text{times } M/36$$

or

$$\begin{bmatrix} -4 & 4 & 0 \\ 4 & -4 & 0 \\ 0 & 0 & -8 \end{bmatrix} \text{times } M/12$$

So adding these three terms we get

$$\begin{bmatrix} -1 + 16 - 4 & 4 & 0 \\ 4 & -1 + 16 - 4 & 0 \\ 0 & 0 & -2 + 32 - 8 \end{bmatrix} \text{times } M/12$$

i.e.

$$\begin{bmatrix} -11 & 4 & 0 \\ 4 & 11 & 0 \\ 0 & 0 & 22 \end{bmatrix} \text{times } M/12$$

So it looks as though the answer to Method 4.1a might be right!

5. *Axes to spin freely.*

By inspection we can see that relative to the centre of mass, the vector $(1, 1, 0)'$ will be an eigenvector, with eigenvalue $15M/12$. (Just multiply it by the matrix to see!)

So the other eigenvectors will be in the directions $(1, -1, 0)'$ and $(0, 0, 1)'$. Their eigenvalues will be $7M/12$ and $22M/12$.

6. *Replacing the missing plate with a point mass.*

With the added mass replacing the missing panel, we have just a 1 metre self-tensor missing from the original 2 metre plate, so the inertia tensor will be

$$\begin{bmatrix} 15 & 0 & 0 \\ 0 & 15 & 0 \\ 0 & 0 & 30 \end{bmatrix} \text{times } M/12$$

i.e.

$$\begin{bmatrix} 5 & 0 & 0 \\ 0 & 5 & 0 \\ 0 & 0 & 10 \end{bmatrix} \text{times } M/4$$

This is not the same as the original inertia tensor!

Exercise 4.5

1. What is the inertia tensor of a thin rod of length $2L$ and mass $2M$?
2. Two such rods are connected to make a cross, aligned with x and y axes respectively. What is its inertia tensor?
3. One of the rods breaks in the middle and the half flies off, leaving a T that is symmetric about the y axis (The negative y axis part has gone). What is the new inertia tensor?
4. About which axes through its centre of gravity can the T spin freely?

Solution

Consider three half-sticks, with centres at $(-0.5, 0, 0)'$, $(0.5, 0, 0)'$, $(0, 0.5, 0)'$.

The centre of mass will be at $(0, 1/6, 0)'$ so we move the coordinates of the masses to $(-1/2, -1/6, 0)'$, $(1/2, -1/6, 0)'$, $(0, 1/3, 0)'$.

Then crank the handle on the $J =$ formula.

$$
\begin{bmatrix}
1/36+1/36+1/9 & -1/12+1/12+0 & 0+0+0 \\
-1/12+1/12+0 & 1/4+1/4+0 & 0+0+0 \\
0+0+0 & 0+0+0 & 1/4+1/36+1/4+1/36+1/9
\end{bmatrix}
$$

i.e.

$$
\begin{bmatrix}
1/6 & 0 & 0 \\
0 & 1/2 & 0 \\
0 & 0 & 2/3
\end{bmatrix}
$$

We add three self-inertia tensors, two of

$$
\begin{bmatrix}
0 & 0 & 0 \\
0 & 1/12 & 0 \\
0 & 0 & 1/12
\end{bmatrix}
$$

and one of

$$
\begin{bmatrix}
1/12 & 0 & 0 \\
0 & 0 & 0 \\
0 & 0 & 1/12
\end{bmatrix}
$$

to get

$$
\begin{bmatrix}
1/6+0+1/12 & 0 & 0 \\
0 & 1/2+1/6+0 & 0 \\
0 & 0 & 2/3+1/6+1/12
\end{bmatrix}
$$

i.e.

$$
\begin{bmatrix}
1/4 & 0 & 0 \\
0 & 2/3 & 0 \\
0 & 0 & 11/12
\end{bmatrix}
$$

Method 4.2
Start with the tensor for the full cross.
Work out the J = tensor for a mass of

4 at $(0, -1/6, 0)$ and a mass of -1 at $(0, 1/3, 0)$

then add the full cross tensor and subtract the half-stick tensor

$$\begin{bmatrix} 1/12 & 0 & 0 \\ 0 & 0 & 0 \\ 0 & 0 & 1/12 \end{bmatrix}$$

to get the same answer as before.

4.9   More About the Tensor

There are some mathematical properties of the inertia tensor that could actually make things clearer!

We have been concerned with the direction of the momentum vector as the body rotates and we are usually keen for it to be balanced, meaning that the momentum vector is aligned with the angular velocity vector. We want $J\omega$ to be in the same direction as ω, which we can express as

$$J\omega = \lambda\omega$$

or

$$(J - \lambda I)\omega = 0 \quad \text{where } I \text{ is the unit matrix.}$$

So we want omega to be an **eigenvector** of **J**.

Every matrix (i.e. tensor) has a set of eigenvectors – though in awkward cases they could be imaginary. Look in the 'eigenvectors' section of www.essdyn.com/maths to see how to find them. But the inertia tensor is rather special. It is symmetric and 'positive semi definite'. That means that whatever axis you spin the body about, it will never have negative kinetic energy.

The three eigenvectors are the **principal axes of inertia** – and they are orthogonal. That means that instead of multiplying the velocity vector by this mysterious tensor, we can resolve the velocity into three components in the directions of the principal axes. We then just multiply each by the corresponding **principal moment** of the inertia tensor and add them.

Chapter 5
Balancing and the Inertia Tensor

Abstract The tensor methods of the previous chapters are used to consider practical balancing problems. Simulations are provided for some rather obviously unbalanced examples, looking at the addition of masses to align the momentum vector with the angular velocity.

5.1 Introduction

For a body to be balanced when it is rotated there are two important requirements. The first is the rather obvious condition that the axis of rotation should pass through the centre of gravity. The second is that the angular momentum vector should be aligned with the axis of rotation.

5.2 Alignment of the Momentum Vector

The angular momentum is given by

$$\mathbf{L} = \mathbf{J}\omega$$

where $\mathbf{J}$ is the inertia tensor and ω is the vector angular velocity.

As you will see, the direction of $\mathbf{L}$ is not necessarily aligned with ω unless $\mathbf{J}$ is diagonal with three equal values.

Let us make up a body with four unit masses, at $(1, 1, 1)'$, $(1, 1, -1)'$, $(-1, -1, 1)$ and $(-1, -1, -1)$.

You can see a picture at www.essdyn.com/sim/4masses.htm.

Now using the expression that we found in the Chap. 4, we have

$$\mathbf{J} = \begin{bmatrix} \sum m(\mathbf{r}_y^2 + \mathbf{r}_z^2) & -\sum m\mathbf{r}_x\mathbf{r}_y & -\sum m\mathbf{r}_x\mathbf{r}_z \\ -\sum m\mathbf{r}_x\mathbf{r}_y & \sum m(\mathbf{r}_x^2 + \mathbf{r}_z^2) & -\sum m\mathbf{r}_y\mathbf{r}_z \\ -\sum m\mathbf{r}_x\mathbf{r}_z & -\sum m\mathbf{r}_y\mathbf{r}_z & \sum m(\mathbf{r}_x^2 + \mathbf{r}_y^2) \end{bmatrix}$$

© Springer International Publishing AG 2018
J. Billingsley, *Essentials of Dynamics and Vibrations*,
DOI 10.1007/978-3-319-56517-0_5

So listing the contributions in turn we have

$$J = \begin{bmatrix} 2+2+2+2 & -1-1-1-1 & -1+1+1-1 \\ -1-1-1-1 & 2+2+2+2 & -1+1+1-1 \\ -1+1+1-1 & -1+1+1-1 & 2+2+2+2 \end{bmatrix}$$

i.e.

$$J = \begin{bmatrix} 8 & -4 & 0 \\ -4 & 8 & 0 \\ 0 & 0 & 8 \end{bmatrix}$$

Now let us consider a rotation about the z axis. We will have momentum

$$L = J \begin{bmatrix} 0 \\ 0 \\ \omega \end{bmatrix}$$

so

$$L = \begin{bmatrix} 0 \\ 0 \\ 8\omega \end{bmatrix}$$

No problems there, as you can see at www.essdyn.com/sim/4massz.htm. But when we rotate it about the x axis it will most certainly wobble! See www.essdyn.com/sim/4massx.htm.

$$L = J \begin{bmatrix} \omega \\ 0 \\ 0 \end{bmatrix}$$

so

$$L = \begin{bmatrix} 8\omega \\ -4\omega \\ 0 \end{bmatrix}$$

which is not aligned with ω.

Exercise 5.1

What happens when we rotate it about the y axis?

5.3 So How Can We Balance It?

We saw that the rotation was balanced about the z axis but not about x and y. Perhaps we can find other axes about which it can spin.

Let us try $\omega = (\omega, \omega, 0)'$. In that case

$$\mathbf{L} = \mathbf{J}\omega$$

will give us

$$\mathbf{L} = \begin{bmatrix} 8 & -4 & 0 \\ -4 & 8 & 0 \\ 0 & 0 & 8 \end{bmatrix} \begin{bmatrix} \omega \\ \omega \\ 0 \end{bmatrix} = \begin{bmatrix} 4\omega \\ 4\omega \\ 0 \end{bmatrix}$$

where the momentum is nicely aligned with the angular velocity.

We can see this in action at www.essdyn.com/sim/4massxy.htm.

But there is another axis about which we can rotate it, still keeping the momentum vector in line with the angular velocity:

If we try $\omega = (\omega, -\omega, 0)'$, we get

$$\mathbf{L} = \begin{bmatrix} 8 & -4 & 0 \\ -4 & 8 & 0 \\ 0 & 0 & 8 \end{bmatrix} \begin{bmatrix} \omega \\ -\omega \\ 0 \end{bmatrix} = \begin{bmatrix} 12\omega \\ -12\omega \\ 0 \end{bmatrix}$$

and again this can be seen in action at www.essdyn.com/sim/4massyx.htm.

It can be shown that any solid body will have three **principal axes** of inertia. These are orthogonal, and can be regarded as an orthogonal set of coordinates. The body might be symmetric about one of the axes, in which case any pair of orthogonal axes that are in turn orthogonal to that one can be used.

5.3.1 Imagine

Think of a coin. A perpendicular axis through its middle, like that of a spinning top, will be the principal axis about which the moment of inertia is greatest. Any two orthogonal diameters can be used as the other axes.

In the case of a sphere, any orthogonal set of axes through its centre will act as principal axes.

A principal axis will be an eigenvector of the inertia tensor. Look at the notes in the www.essdyn.com/maths site to see how to find it.

Instead of modifying the solid to achieve balance, you can find a slightly different axis to spin about, first rotating the object until one of its principal axes lines up with the spin axis. This is done in a vertical axis **spin-dryer**. With a cunning arrangement of springs the machine can be made to pull into alignment automatically. Getting those dynamics right might be a topic to consider in a more advanced course!

5.4 Balancing by Adding Masses

You will be familiar with wheel-balancing, where lead weights are added to a motor car's wheel rim. The first requirement is to bring the centre of gravity of the wheel to lie on the line of the axle. Next, however, a principal axis of the wheel must be aligned with the axle by the addition of more weights.

What is actually happening is that a second system is being added to the first. The unbalanced wheel had an inertia tensor with some non-zero off-axis terms.

The balancing weights form a system that also has non-zero off-axis terms, that are equal in magnitude but opposite in sign to those of the wheel. When the two systems are added together, the result is a system with no off-axis non-zero terms.

Exercise 5.2
Look back at Exercise 4.1 part 5 of the Inertia module. What did you get for the inertia tensor?

Then you considered masses at $(1, -1, 0)$ and $(-1, 1, 0)$. What do you get when you add these two systems together? You should have seen them combine to form a balanced system.

So now consider the following problem.

Starting with our unbalanced four mass example, with four unit masses, at $(1, 1, 1)'$, $(1, 1, -1)'$, $(-1, -1, 1)$ and $(-1, -1, -1)$, how can you add just two unit masses to balance it? (See www.essdyn.com/sim/4massx.htm)

They must not be placed further out along the x-axis than the other masses.

Solution
In Section 2, we found the inertia tensor to be

$$\mathbf{J} = \begin{bmatrix} 8 & -4 & 0 \\ -4 & 8 & 0 \\ 0 & 0 & 8 \end{bmatrix}$$

so our balancing masses must cancel the -4 term without introducing any other off-axis terms.

Since the centre of mass is at the origin, our added masses must be symmetrical. It seems reasonable to put them at $(1, y, z)$ and $(-1, -y, -z)$.

Our added masses will have inertia tensor

$$\begin{bmatrix} 2(y^2 + z^2) & -2y & -2z \\ -2y & 2(z^2 + 1) & -2yz \\ -2z & -2yz & 2(y^2 + 1) \end{bmatrix}$$

so we need

$$-2y = 4$$
$$-2z = 0$$
$$-2yz = 0$$

Our masses are at $(1, -2, 0)$ and $(-1, 2, 0)$.

Chapter 6
Couples, Moments and Euler's Equations

Abstract There is a proof given here of Euler's equations, but you can really just take them on trust. They tell you that if the principal moments of inertia are all different, the equations become non-linear, with the rate-of-change of each component of the angular velocity depending on the product of the other two. Solving such equations is very difficult indeed, but setting up a simulation to represent them is no problem at all, as you can see in the simulation of the 'dancing T-handle'. If two of the principal moments are equal, everything becomes much simpler. One of the rates-of-change becomes zero, so the angular velocity about that axis is constant and the equations become linear once again. That describes the case of the gyroscope, where its spin is assumed to remain constant (even though it will actually slow down unless powered). We can add in any couple that is applied to it and get equations for its precession. Another way to think of the gyroscope is as a carrier of a huge angular momentum about its spin axis. When an orthogonal couple C is applied, momentum C.dt is added to that momentum for each small time-step dt. The new momentum vector is the sum of the original vector plus the small C.dt at right-angles to it, so its direction will be rotated slightly. The result is the precession. But the gyroscope has a chapter all to itself.

6.1 Tossing a Coin

This chapter is probably easier to understand if we consider a practical example.

When a coin is tossed to decide a decision, it usually spins about one of its diameters. This is one of the principal axes and we have no trouble in predicting the motion – apart from deciding whether it will land with 'heads' or 'tails'.

But if you want to throw a coin as far as possible, you will probably spin it 'Frisbee' style, rotating like an athlete's discus. Now it is spinning about an axis perpendicular to the coin and once again the motion can be solved easily.

However if the coin starts its motion spinning about an axis midway between these two it will 'tumble' in a way that is hard to solve by any method apart from simulation. Try it. Flick a coin that is tilted between finger and thumb and watch its motion as it falls.

© Springer International Publishing AG 2018

J. Billingsley, *Essentials of Dynamics and Vibrations*,

DOI 10.1007/978-3-319-56517-0_6

Now imagine the task of computing a rotational thrust to stabilise a tumbling satellite.

6.2 Rate of Change of Momentum

We have to find a relationship between the applied couple (or torque) and the rate of change of the angular momentum. This is simply

$$C = \frac{d}{dt}(J\omega).$$

(In my notation I have used C for couple, but you will often find the Greek letter tau for torque.)

It starts to look less simple when we realise that both omega and J will be changing with time.

$$C = J\dot{\omega} + \dot{J}\omega$$

The rate of change of omega is simple enough, but as the body rotates its inertia tensor is transformed by

$$TJ_0T^{-1}$$

where J_0 is the diagonal matrix of principal moments and T is the transformation that brings the body to its present position.

So how is T related to the angular velocity? One thing is certain. We cannot simply integrate omega to get values corresponding to pitch, roll and yaw. All looks well when we rotate the angles one at a time, but to see what happens when we change two at a time try manipulating the simulation at www.essdyn.com/sim/rollyaw.htm.

Keep your eye on any particular ball and observe that when both roll and yaw are changed together it does not rotate in a circle – or even in an ellipse given by the perspective view of a circle. Instead it looks like a fairground ride.

You should perhaps read the web maths help on **Orientation** again to see the relationship between the rotation angles and the matrix transformation. You will see that the relationship between them is highly nonlinear.

So how did the simulation of oblique rotations work in the simulation at www.essdyn.com/sim/4massxy.htm?

In a small time increment ∂t, the system will rotate through an angle

$$\begin{bmatrix} \omega_x \delta t \\ \omega_y \delta t \\ \omega_z \delta t \end{bmatrix}$$

The transformation $\mathbf{T}$ will therefore be multiplied by a rotation transform

$$
\begin{bmatrix}
1 & 0 & 0 \\
0 & \cos\omega_x\delta t & -\sin\omega_x\delta t \\
0 & \sin\omega_x\delta t & \cos\omega_x\delta t
\end{bmatrix}
\begin{bmatrix}
\cos\omega_y\delta t & 0 & \sin\omega_y\delta t \\
0 & 1 & 0 \\
-\sin\omega_y\delta t & 0 & \cos\omega_y\delta t
\end{bmatrix}
\begin{bmatrix}
\cos\omega_y\delta t & -\sin\omega_z\delta t & 0 \\
\sin\omega_z\delta t & \cos\omega_z\delta t & 0 \\
0 & 0 & 1
\end{bmatrix}
$$

In the limit, for very small ∂t this becomes

$$
\begin{bmatrix}
1 & -\omega_z\delta t & \omega_y\delta t \\
\omega_z\delta t & 1 & -\omega_x\delta t \\
-\omega_y\delta t & \omega_x\delta t & 1
\end{bmatrix}
=
\begin{bmatrix}
1 & 0 & 0 \\
0 & 1 & 0 \\
0 & 0 & 1
\end{bmatrix}
+
\begin{bmatrix}
0 & -\omega_z & \omega_y \\
\omega_z & 0 & -\omega_x \\
-\omega_y & \omega_x & 0
\end{bmatrix}\delta t
$$

so

$$
\frac{\mathrm{d}\mathbf{T}}{\mathrm{d}t}
=
\begin{bmatrix}
0 & -\omega_z & \omega_y \\
\omega_z & 0 & -\omega_x \\
-\omega_y & \omega_x & 0
\end{bmatrix}\mathbf{T}
$$

This is very useful for animating the simulation, just as in the examples we have seen. But it does not help us in solving the equations for omega.

Let us try another approach.

6.3 Euler's Equations

Instead of taking an absolute frame of reference, let us use a set of axes that are moving with the body. Then its inertia tensor will not change. In fact, we can take the principal axes as the frame of reference, so that the inertia tensor simply becomes

$$
\mathbf{J}=
\begin{bmatrix}
J_x & 0 & 0 \\
0 & J_y & 0 \\
0 & 0 & J_z
\end{bmatrix}
$$

As you can see in Appendix 1 ('Mathematicians and Operators'), we can relate rotating axes to fixed axes by applying the operator

$$
\frac{\mathrm{d}}{\mathrm{d}t} = \frac{\partial}{\partial t} + \omega \times
$$

so the applied couple $\mathbf{C}$ is given by

$$
\mathbf{C} = \frac{\partial(\mathbf{J}\omega)}{\partial t} + \omega \times (\mathbf{J}\omega)
$$

i.e.

$$\mathbf{C} = \begin{bmatrix} J_x\dot{\omega}_x \\ J_y\dot{\omega}_y \\ J_z\dot{\omega}_z \end{bmatrix} + \omega \times \begin{bmatrix} J_x\omega_x \\ J_y\omega_y \\ J_z\omega_z \end{bmatrix}$$

(since J is constant and diagonal)

which expands to three equations for the three components

$$C_x = J_x\dot{\omega}_x + (J_z - J_y)\omega_y\omega_z$$

$$C_y = J_y\dot{\omega}_y + (J_x - J_z)\omega_x\omega_z$$

$$C_z = J_z\dot{\omega}_z + (J_y - J_x)\omega_x\omega_y$$

These are Euler's equations. (Again I remind you that you will often see them with Greek letter tau (for torque) instead of $\mathbf{C}$.)

These are somewhat complicated because $\mathbf{C}$ is represented in terms of the instantaneous coordinates of the principal axes of the body. But of course if you are applying corrections to a spacecraft from within the spacecraft, that is exactly what you need. Your attitude control jets will be fixed within the spacecraft.

If we are only interested in the free motion of the body with no applied couple, there is no problem anyway. So in this case we set $\mathbf{C}$ to zero and get a set of differential equations for the rate of change of omega

$$\dot{\omega}_x = \frac{J_y - J_z}{J_x}\omega_y\omega_z$$

$$\dot{\omega}_y = \frac{J_z - J_x}{J_y}\omega_x\omega_z$$

$$\dot{\omega}_z = \frac{J_x - J_y}{J_z}\omega_x\omega_y$$

6.4   Simulating the Coin

Now we have all that we need. We have state variables that are the three components of omega, plus the transformation matrix T that has actually only three degrees of freedom.

The procedure for simulating the coin is as follows:

1. Set up the initial conditions.
2. Using the present values of the three components of omega, calculate their rate of change.

3. Update omega by adding dt times the rate of change.
4. Update T by multiplying it by (I + skew(omega)dt), where skew(omega) is the matrix we found at the end of Section 2.
5. Use T to transform the orientation of the coin and display it.

Now you could investigate the linked simulation at www.essdyn.com/sim/coin. htm and add some of your own code to it.

Exercise 6.1
Follow the link www.essdyn.com/sim/domino.htm to the simulation of a tumbling domino.

Play with it by trying different values for the initial angular velocity.

The masses in this simulation are at (0.3, 0.5, 0), (0.3, −0.5, 0), (−0.3, 0.5, 0) and (−0.3, −0.5, 0), ensuring that the principal moments are all different.

To change the simulation to that of a coin, edit the code in the window to represent a system with masses at (0.5, 0.5, 0), (0.5, −0.5, 0), (−0.5, 0.5, 0) and (−0.5, −0.5, 0).

The code will calculate new values for the principal moments of inertia.

What do you notice about the displayed value of wz as the balls move? Why?

6.5 The Dancing T-handle

There is some fascinating video at https://www.youtube.com/watch?v=1n-HMSCDYtM.

A small tool spinning in zero gravity flips to and fro.

But it can all be explained by Euler's equations!

Look at www.essdyn.com/sim/t-handle.htm to see the simulation in action!

It tells us that an object can rotate steadily about any of its principal axes. But add a very slight component about one of the other axes and the rotation can build up a 'wobble' that can cause the orientation to 'flip'. By playing with the simulation you can conduct some simple research:

• Will this sort of instability apply to any of the principal axes?
• How does the 'time to flip' relate to the initial off-axis component?

Chapter 7
Gyroscopes

Abstract The gyroscope can be a fascinating toy. You can buy one on the Internet for a few dollars. It gives a practical appreciation of angular momentum. The simulations throughout this book are intended to help you to visualise some difficult concepts. However those simulations cannot let you feel the torque of a real gyroscope. When you change the angle of the axis, you will feel torque about an axis perpendicular to it. The gyroscope will seem to be trying to twist out of your hands. If the gyroscope is suspended from a string, attached to one end of the axis, the axis will remain horizontal while the gyroscope turns steadily about a vertical axis. The gyroscope is 'precessing'. As well as precessing, gyroscopes can perform 'nutation', cyclic 'wobbling' of the axis. But that will not be considered here. If you are curious about the subject, you might like to look for the origin of the saying, 'Sleeping like a top.'

7.1 Introduction

Some years ago gyroscopes and spinning tops could be found in any toy box that contained Meccano pieces. Now you can source some online. They are excellent for giving practical experience on the difference between a force and a couple.

The essence of a gyroscope is a spinning wheel, which therefore has a large angular momentum vector that lies along its axis. To rotate the axis sideways will cause this vector to change, requiring a substantial couple to give any rate of rotation.

This tendency to stay aligned made gyroscopes very useful as sensors for stabilising aircraft and spaceships, though today the sensors are more likely to use microminiature 'tuning forks'.

At http://www.youtube.com/watch?v=cquvA_IpEsA there is a YouTube video at which will show you a toy gyroscope in action, but much of the narrative is WRONG! The narrator refers to 'forces' balancing the gyroscope when he should really refer to 'couples'. That is exactly the mistake that cost Eric Laithwaite his reputation.

© Springer International Publishing AG 2018
J. Billingsley, *Essentials of Dynamics and Vibrations*,
DOI 10.1007/978-3-319-56517-0_7

Fig 7.1 Gyroscope

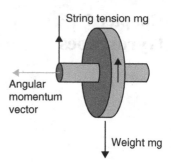

Laithwaite presented a Christmas lecture at the Royal Society which you can see at http://www.gyroscopes.org/1974lecture.asp (though some large downloads are involved). In it he cast doubts on Newton's laws.

The thing that seems to have most impressed him is that you can suspend the gyroscope on a string from one end of the axle and the axle will remain horizontal while the string appears to stay vertical. You can see this in action at http://www. youtube.com/watch?v=8H98BgRzpOM&feature=related.

But you will see that the gyroscope is precessing, rotating about the vertical string, and as it does this its angular momentum is changing. This rate-of-change is the result of the **couple** formed by the vertical tension in the string and the equal and opposite weight of the gyroscope acting through its centre of gravity.

The equal and opposite forces of string tension and weight form a couple, represented by a vector 'into the page'. In each small time dt, this couple times dt will be added to the momentum vector causing it to rotate 'into the page' also. So rather surprisingly the gyroscope does not rotate around the axis of the applied couple, but rotates about the vertical axis of the string. The string continues to hang vertically.

As you can see from Fig. 7.1, that rotation, called 'precession', will cause the body of the gyroscope to move towards us and then continue to circle the string.

In the following sections we will analyse the motion.

7.2 Euler's Equations Again

We have

$$C_x = J_x\dot{\omega}_x + (J_z - J_y)\omega_y\omega_z$$
$$C_y = J_y\dot{\omega}_y + (J_x - J_z)\omega_x\omega_z$$
$$C_z = J_z\dot{\omega}_z + (J_y - J_x)\omega_x\omega_y$$

where the components of the couple are represented relative to the principal axes of the body.

If we spin the gyroscope wheel about the x axis, we see that J_y and J_z will be equal because of symmetry. That means that without any applied couple, ω_x will remain constant. You should have noticed this when you modified the code of the 'spinning domino' at www.essdyn.com/sim/domino.htm.

These equations will lead us to a simulation, just as before with the **spinning coin**, but there are some more problems along the way. For the coin and the spinning domino, we considered the case where there was no applied torque. In the case of the gyroscope, we want to apply the torque resulting from the weight of the rotor and the equal and opposite tension in the string.

This torque has to rotate with the gyroscope axis for any simulation to be representative.

7.3 Rotating the Couple

The simulation at www.essdyn.com/sim/precess.htm is simplified by representing the rotor as a square of four point-mass balls, numbered 0 to 3 in the code. It includes two black balls, 4 and 5, that will represent the spin axis. These have not been included in the calculation of the inertia tensor and therefore have zero effective mass. However when we run the transformation to calculate the positions of the balls, we will have found the coordinates of the ends of the spin axis and can use these to calculate the couple to apply.

The spin is much slower than that of a real gyroscope, otherwise the picture would just be a blur. But the gravitational torque is set to a much smaller value, to give a reasonable rate of precession.

In this simulation we have the z-axis 'out of the page' and the y-axis upwards. The direction of the couple in world axes will thus be the cross product of the coordinates of ball 4, at an end of the axis, with $(0, -g, 0)$, the vector weight.

```
cworld=Array(ballt[4][2],0,-ballt[4][0]);
```

We transform this back to gyro axes by multiplying it by the transpose of the transformation matrix – which is also its inverse.

Run the **precession** simulation to see it in action, and you might notice several other features. One of the problems of discrete-time simulation is the approximation implied in 'Euler integration'. (Yes, it's the same Euler again!) We are adding dt times the rate of change to approximate to an integral.

Consider an example where you are running in a small circle. Each step you take is in the direction of the tangent to the circle – but if the step is not very small it will take you to a point outside the circle. Take another step in the direction of a tangent to a new circle through that point and you will move further out still. Soon the small circle has become a large circle that is growing exponentially.

In the same way, if you watch the simulation for a while you will see the 'wobble' build up in a way that represents instability in the calculation, though not necessarily in the gyroscope. Double the value of dt and you will see it build up much faster, halve it and the simulation will be more accurate.

7.4 The Gyroscope as a Top

It is when the gyroscope acts as a top that we see the most interesting behaviour. In Fig. 7.1 we can see the gyroscope supported by a string at the end of the axle. A top spinning on a sharp point fixes that suspension point and we can usefully take it as the origin of our coordinate system.

We can represent the angle of the axis by two parameters, θ and ψ. θ is the inclination of the axis from the vertical z axis, while ψ is a horizontal rotation. We can denote a unit vector $\mathbf{r}$ in this direction as

$$\mathbf{r} = (\cos\psi \, \sin\theta, \, \sin\psi \, \sin\theta, \, \cos\theta)'.$$

If the top is spinning at N rad/s and the principal moment of inertia is the scalar J_1, the momentum about the spin axis is NJ_1. The major component of the momentum vector $\mathbf{C}$ will be NJ_1 times $\mathbf{r}$. But there will be an additional component $J_2 \, \mathrm{d}\mathbf{r}/\mathrm{d}t$, where J_2 is the minor moment of inertia.

Meanwhile the torque due to gravity will be

$$\tau = mh\mathbf{r} \times (0, 0, -g)' = (-mgh \cos\psi \sin\theta, mgh \sin\psi \sin\theta, 0)'$$

where h is the height of the centre of mass.

By expanding

$$\mathrm{d}\mathbf{C}/\mathrm{d}t = \tau$$

we get three equations for the components of d$\mathbf{r}$/dt.

The result is a set of equations describing both precession and nutation, a cyclic 'wobble' as the top precesses.

Now in a later chapter you will meet 'modes', where one vibration is superimposed upon another. For small perturbations we can consider modes here.

One solution of the equations will be a steady precession, where the 'latitude' remains constant while the 'longitude' rotates at a constant rate so that the torque due to gravity is equal to the rate-of-change of the horizontal component of the spin momentum

$$J_1 N \sin\theta \, \mathrm{d}\psi/\mathrm{d}t = mgh \sin\theta$$

So the precession rate will be independent of the angle that the top leans from the vertical.

'Nutation' will add a second mode to this one, where the axis makes small cyclic excursions around this steady precession. Intuitively it can be seen as follows:

Velocity of the axis 'north' will cause a couple that integrates to give the axis an additional velocity 'west'. But this velocity in turn gives a change in couple that integrates to give a velocity 'south'. And round it goes!

We have the two first-order equations that we need to set up an oscillation, as we will find in the Chap. 11. But the details are left to you.

You can see precession and nutation in action at www.essdyn.com/sim/top.htm.

You can see Eric Laithwaite's obituary at http://www.independent.co.uk/news/obituaries/obituary-professor-eric-laithwaite-1288502.html   and   if   you   enter 'Laithwaite gyroscope' into your search engine you will see many reports of lectures and conjecture on space propulsion.

Chapter 8
Kinematics

Abstract One of the more useful applications of the matrix techniques that you have been learning is to the control of a robot. Most industrial robots consist of a chain of 'revolute' axes which we can think of waist, shoulder, elbow, wrist and so on. But the position of the 'end effector', the hand that does the work, is a somewhat complicated combination of the functions that depend on all these angles.

By now you should be familiar with the three-by-three matrices

$$\begin{bmatrix} 1 & 0 & 0 \\ 0 & c & -s \\ 0 & s & c \end{bmatrix}$$

$$\begin{bmatrix} c & 0 & s \\ 0 & 1 & 0 \\ -s & 0 & c \end{bmatrix}$$

and

$$\begin{bmatrix} c & -s & 0 \\ s & c & 0 \\ 0 & 0 & 1 \end{bmatrix}$$

that define rotations about the x, y and z axes respectively.

But now we need to combine translations with these rotations, so that we can 'move down' the parts of the robot, from shoulder to elbow, say.

We could simply add the displacement to our present coordinate, but we would really like something that can be applied using the standard computer matrix multiplication routine.

© Springer International Publishing AG 2018
J. Billingsley, *Essentials of Dynamics and Vibrations*,
DOI 10.1007/978-3-319-56517-0_8

So we 'fatten up' the matrix to become four-by-four and add a fourth component, which is always 1, to our position vector to become $(x, y, z, 1)'$. Now

$$\begin{bmatrix} c & 0 & s & L \\ 0 & 1 & 0 & 0 \\ -s & 0 & c & 0 \\ 0 & 0 & 0 & 1 \end{bmatrix}$$

will represent a rotation about the y axis combined with a translation L in the x direction. But does the translation happen before or after the rotation when we are describing the kinematics of a robot arm?

This chapter tries to remind you of the theory, while the next will put it into practice.

8.1 Introduction

When a robot has enough joints, the gripper endpoint can move in three dimensions. But the end-effector has three more rotational degrees of freedom. We must revise some of the mathematics of three-dimensional motion to find the tools to analyse this motion.

You will find that the notes in the '**Vectors and transformations**' help at www.essdyn.com/maths cover much of the same ground as is covered below. You will find it helpful to read both versions.

8.2 Coordinates and Transformations

A point P in space is defined by a three dimensional vector, but the way to represent it is not unique. The most obvious form is Cartesian, the three coordinates being found by resolving the vector from the origin in the directions of three orthogonal vectors through that origin. There are also spherical polar coordinates, equivalent to defining the latitude, longitude and distance of the point from the origin, also cylindrical polar in which the point is represented by radius, direction and height.

Not only is the location of the origin a matter of choice, we can orient the orthogonal vectors of Cartesian coordinates with three more degrees of freedom.

For now, however, let us take it that the origin is fixed and that we have three unit vectors $\underline{i}, \underline{j}$ and $\underline{k}$ defining the x, y and z directions.

Our point P can be represented as $(x, y, z)'$ meaning

$$x\,\underline{i} + y\,\underline{j} + z\,\underline{k}$$

As we saw in Chap. 2, when the point moves, x, y and z will vary as functions of time. Now we can take the derivatives of the vector components to calculate the velocity and acceleration. It is worth making a few remarks about these.

As it moves, P will follow a curve in space. The velocity vector will be a tangent to this curve at P. The acceleration can be broken into two perpendicular components. One of these is in the same direction as the velocity, representing a change in speed, while the other is perpendicular to the path, aligned through the instantaneous centre of rotation, the centre of curvature of the path at that point.

This may seem too simple – and it is.

When we start to analyse the motion of a robot, we must deal with six dimensions, not just three. We are concerned with solid bodies, not mere points in space. We have three dimensions of freedom in the location of one particular point of the object, but then we can perform three rotations to orient the object in space. We might think of these rotations as movement about the pitch, roll and yaw axes of an aircraft.

Instead of the vector coordinates of just one of its points, we have to think of the position and orientation of the object as being defined by the transformation that maps each of its points to the new position that it takes up. Let us first consider the transformation of rotation.

8.3 Rotations

We can set up a coordinate system of three orthogonal axes in the object. To start with, these will coincide with our 'reference system' axes $\underline{i}$, $\underline{j}$ and $\underline{k}$ that stay fixed. But as we rotate the object about the origin, its axes will move to be three other orthogonal vectors through the origin.

Let us consider three such unit vectors $\underline{a}$, $\underline{b}$ and $\underline{c}$, passing through the origin of our coordinate system and orthogonal to each other.

A point expressed in terms of these vectors as coordinates $(x, y, z)'$ will be

$$\underline{a}\,x + \underline{b}\,y + \underline{c}\,z$$

This can be expanded as

$$(a_1\underline{i} + a_2\underline{j} + a_3\underline{k})x + (b_1\underline{i} + b_2\underline{j} + b_3\underline{k})y + (c_1\underline{i} + c_2\underline{j} + c_3\underline{k})z$$

or

$$\begin{bmatrix} \underline{i} & \underline{j} & \underline{k} \end{bmatrix} \begin{bmatrix} a_1 & b_1 & c_1 \\ a_2 & b_2 & c_2 \\ a_3 & b_3 & c_3 \end{bmatrix} \begin{bmatrix} x \\ y \\ z \end{bmatrix}$$

to give the coordinates of the same point in terms of the reference system.

(Remember that a vector makes a mixture of the columns of a matrix that it multiplies – see the **'matrices' notes** in the maths help site.)

We transform the coordinates to the reference axes by multiplying $(x, y, z)'$ by this matrix A. So let us look at some of the properties of A.

Since they are unit vectors, $\underline{a}.\underline{a} = 1$, $\underline{b}.\underline{b} = 1$ and $\underline{c}.\underline{c} = 1$

Also, since the vectors are orthogonal, the scalar product of any two different vectors is zero, e.g. $\underline{a}.\underline{b} = 0$.

Let us consider the product of A with its transpose:

$$AA' = \begin{bmatrix} a_1 & a_2 & a_3 \\ b_1 & b_2 & b_3 \\ c_1 & c_2 & c_3 \end{bmatrix} \begin{bmatrix} a_1 & b_1 & c_1 \\ a_2 & b_2 & c_2 \\ a_3 & b_3 & c_3 \end{bmatrix}$$

Remember the 'scalar products' way to look at matrix multiplication. We see that

$$AA' = \begin{bmatrix} a\cdot a & a\cdot b & a\cdot c \\ b\cdot a & b\cdot b & b\cdot c \\ c\cdot a & c\cdot b & c\cdot c \end{bmatrix}$$

But from what we know of these scalar products

$$AA' = \begin{bmatrix} 1 & 0 & 0 \\ 0 & 1 & 0 \\ 0 & 0 & 1 \end{bmatrix}$$

So

$$A'A = I$$

or

$$A' = A^{-1}$$

A rotation transformation matrix is extremely easy to invert!

There is a further property that we have to preserve; the axes must make up a 'right handed' set. The conventional set of axes will be $\underline{i}$ and $\underline{j}$, as we draw x and y on a horizontal sheet of graph paper, and $\underline{k}$ vertically upwards in the 'z' direction.

Because it reverses the x coordinate, the matrix

$$\begin{bmatrix} -1 & 0 & 0 \\ 0 & 1 & 0 \\ 0 & 0 & 1 \end{bmatrix}$$

would map a left-hand glove into a right-hand glove, something no rotation could do. Yet it satisfies the property of having three mutually orthogonal unit vectors as its rows and its columns. What is wrong?

A property of a rotation is that there is an axis about which the rotation takes place. Now if a vector $\underline{\xi}$ is aligned with this axis, it is not changed by being transformed by A, in other words

$$A\,\underline{\xi} = \underline{\xi}$$

So $\underline{\xi}$ is an eigenvector of A, with eigenvalue 1. All the eigenvalues of a rotation must be 1, so the determinant of A must be 1. The determinant of the glove-bending matrix is -1, so it cannot represent a rotation.

8.4 Examples of Rotations

We should look at some special examples of rotation matrices.

If we rotate the $x\,y$ plane by an angle θ_1 about the z axis, we get new coordinates

$$(x \cos \theta_1 - y \sin \theta_1, x \sin \theta_1 + y \cos \theta_1, z)'$$

The z component stays the same.

In matrix terms, the transformation is

$$\begin{bmatrix} \cos \theta_1 & -\sin \theta_1 & 0 \\ \sin \theta_1 & \cos \theta_1 & 0 \\ 0 & 0 & 1 \end{bmatrix}$$

A rotation θ_2 about the y axis would be represented by

$$\begin{bmatrix} \cos \theta_2 & 0 & \sin \theta_2 \\ 0 & 1 & 0 \\ -\sin \theta_2 & 0 & \cos \theta_2 \end{bmatrix}$$

Note that a positive rotation is 'clockwise looking out along the axis', so this tips the x axis down.

If we multiply the matrices together to get the result of applying both transformations, we will start to build up a string of sines and cosines that will be lengthy to write and muddling to read. Therefore we use considerable abbreviation and write $\cos \theta_1$ as c_1, $\sin \theta_1$ as s_1 and so on.

If we apply these in order, the transformed coordinates will be

$$
\begin{bmatrix} c_2 & 0 & s_2 \\ 0 & 1 & 0 \\ -s_2 & 0 & c_2 \end{bmatrix}
\begin{bmatrix} c_1 & -s_1 & 0 \\ s_1 & c_1 & 0 \\ 0 & 0 & 1 \end{bmatrix}
\begin{bmatrix} x \\ y \\ z \end{bmatrix}
$$

Note that the transformation that is applied first is closest to the vector, in other words the matrices are ordered right to left. Note too that the order is important and must not be changed. Here the result is

$$
\begin{bmatrix} c_1 c_2 & -s_1 c_2 & s_2 \\ s_1 & c_1 & 0 \\ -c_1 s_2 & s_1 s_2 & c_2 \end{bmatrix}
\begin{bmatrix} x \\ y \\ z \end{bmatrix}
$$

Check that the columns are unit vectors that are mutually orthogonal.

You can see the product of rotations about the axes making a general rotation matrix at www.essdyn.com/sim/3Dmatdeg.htm.

So far we have been considering transformations that leave the origin fixed, but we must also be able to move the coordinates anywhere in three dimensions.

8.5 Translations

To move an object a vector distance, we simply add that vector to every one of its points.

For example a point $(x, y, z)'$ can be moved a vector distance $(1, 2, 3)'$ to arrive at $(x + 1, y + 2, z + 3)'$ – it's not really difficult! To find the new vector we simply add the displacement to it.

The problem is that we now have two different processes for dealing with the two types of movement, rotation and translation. One involves multiplying the point coordinates by a three-by-three matrix while the other involves adding constants to each component. Can we find some way of gluing them together into a single operation? If we can, we can start to deal with combinations of transformations, such as 'screwing' where the object is rotated at the same time as being moved along the rotation axis.

8.6 Combining Rotations and Translations

We have to appease the mathematicians! Rotation is a transformation given by a simple multiplication of a vector by a matrix, but the ability to add a constant to the result requires an 'affine' transformation.

However there is a way around the problem.

Suppose that instead of writing our vector as $(x, y, z)'$ we write it as $(x, y, z, 1)'$.

What is the 1 for? It gives something for a matrix to grab onto to add a translation $\underline{d}$ to the vector! But now the vector has four components and the matrix is four by four.

We can 'partition' a matrix to see its various parts in action, so if we write $T\underline{x}$ for the product of our point with a transformation matrix, now four by four, we can break it down as follows.

$$\begin{bmatrix} A & \underline{d} \\ 0\,0\,0 & 1 \end{bmatrix} \begin{bmatrix} \underline{x} \\ 1 \end{bmatrix} = \begin{bmatrix} A\underline{x} + \underline{d} \\ 1 \end{bmatrix}$$

So at the expense of changing our matrices to four by four, where the bottom row is always $(0, 0, 0, 1)$, we can apply any combination of rotations and translations, just by multiplying the transformation matrices together.

8.7   Rotations, Translations and Screws

If we multiply our vector augmented by the extra '1' by the combination of rotation R and displacement $\underline{d}$

$$\begin{bmatrix} R & \underline{d} \\ 0\,0\,0 & 1 \end{bmatrix} \begin{bmatrix} \underline{x} \\ 1 \end{bmatrix} = \begin{bmatrix} R\underline{x} + \underline{d} \\ 1 \end{bmatrix}$$

Much depends on the relationship between the translation vector $\underline{d}$ and the axis of rotation.

Now if $\underline{d}$ is zero, we have a simple rotation about the origin.

If the displacement $\underline{d}$ is in the same direction as the axis of rotation, we have what is called a 'screw' – for obvious reasons. The object rotates about the axis as it moves along it.

But in general, the displacement will have a component in a plane orthogonal to the axis. As we saw in Chap. 2, that component can be combined with the rotation to represent a rotation about an axis that does NOT pass through the origin

This transformation T takes the origin to the coordinates $\underline{d}$. So how would we find out if T represents a rotation or a screw?

How would we find the axis of rotation?

Suppose that $\underline{c}$ is a point on the axis of rotation. What happens when we apply the transformation?

If the transformation is a rotation (not a screw) then $\underline{c}$ stays in exactly the same place. We have

$$R\underline{c} + \underline{d} = \underline{c}$$

or

$$(R - I)\underline{c} = -\underline{d}$$

Although this is a set of three simultaneous equations in the three coordinates of $\underline{c}$, we cannot solve the equations to find a unique point $\underline{c}$. There are an infinity of points on the axis, and these equations could lead to any of them.

To nail the solution down to one specific point we have to add an extra condition, such as that the z-component of $\underline{c}$ must be zero –'Where does the rotation axis cut the x-y plane?'

Now that there are just two unknowns, the three equations must reduce to two – the third equation must be 'consistent' with the other two. Either the third equation amounts to a mixture of the first two equations, or the first two equations are equivalent.

But if this is not the case, T is a screw.

In this case, there is NO solution to our set of equations, since every point c on the axis of rotation will be transformed into another point 'further along'. There will be no point that remains in the same place.

Since $(R - I)$ is singular its determinant must be zero. (It has to be singular, since if applied to an eigenvector of R the result is a vector of zeros.)

To test for consistency, drop in the column vector $\underline{d}$ in place of one of the columns of $(R - I)$. If the determinant is not again zero, the transformation is not a rotation but a screw.

8.8 Rotations and Screws – A Bit More Explanation

The transformation

$$R\underline{x}$$

changes the location of the point $\underline{x}$ by a vector amount

$$R\underline{x} - \underline{x}$$

or

$$(R - I)\underline{x}$$

Since R is a rotation, it will move any point that is not on the axis of rotation in a direction that lies in a plane that is orthogonal to that axis.

Now this displacement vector is a combination of the vectors of the columns of $(R - I)$, which lie in a plane.

For $R\underline{x} + \underline{d}$ also to represent a rotation, $\underline{d}$ must lie in the same planar direction. The determinant of three coplanar vectors is zero, so if we substitute $\underline{d}$ for one of the columns of $(R - I)$ the resulting determinant must be zero.

If that determinant is not zero, then the transformation is a screw.

Chapter 9
Kinematic Chains

Abstract Now it is time to string four-by-four matrices together, so that we can compute the position and orientation of the end-effector in terms of the axis angles.

To avoid confusion we can separate out each individual action, such as an individual rotation or an individual translation down a limb, so that they are only combined when we multiply their matrices together.

We have to be clear about the order in which to put the matrices. If we start at the 'hand', a point with coordinates $(x, y, z, 1)'$ relative to the hand would have coordinates $(x + L, y, z, 1)'$ relative to an elbow, if the hand is just a translation L in the x direction from it.

So if we think of 'travelling back from the hand', the matrices will build up from right to left.

If we think of 'travelling out from the base', they will build up from left to right. So we might end up with a product of:

(waist rotate)(shoulder rotate)(translate shoulder to elbow)(elbow rotate) (translate elbow to wrist)(wrist rotate) – and so on, in that order.

But when you come to the grind of multiplying the matrices together, as long as you keep them in the right order you can pair them off in any way that you wish, starting from the right or left or even in the middle.

This is one of the more straightforward tasks. If you are designing a robot controller, you will have the problem of 'inverse kinematics', of calculating a set of axis angles to give some desired position and orientation of the end-effector. But that is another story.

9.1   Introduction

Many robots take the form of a 'chain' of links, just as our shoulder connects the upper arm, then the elbow connects the forearm and the wrist connects the hand. We can analyse the motion by considering a chain of transformations, multiplying them together in the right order.

© Springer International Publishing AG 2018   83
J. Billingsley, *Essentials of Dynamics and Vibrations*,
DOI 10.1007/978-3-319-56517-0_9

There are other robots, like the 'Stewart platform' where there are loops of links, but we will not discuss them here.

9.2 Open Chains

The most usual form for a robot is a chain of links with actuated joints between them. These joints can be 'revolute', a sort of powered hinge, or 'prismatic' with one member sliding past another. We will refer to both types as 'axes'. Although some kinematic chains can be 'closed', such as the four-bar linkage of Fig. 9.1, most robots are 'open' where only one end of the chain is fixed. (A small proportion of writers refer to this mechanism as a three-bar linkage. There are indeed three bars, but the line joining the base pivots form a virtual bar.)

Fig. 9.1 Four-bar linkage

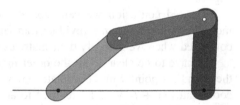

When we consider the tool-piece of a robot, its location in space has been transformed by the motion of every axis in turn that moves it. Before we can address the task of deciding on joint angles or displacements to put the tool where we want it, we have to derive an expression for its location and orientation in terms of the joint axis variables.

9.3 Chains of Axes – Puma Example

When we have just one movable axis, there is a single transformation and all is straightforward. When we have a robot such as the Unimation Puma, with six degrees of freedom, we have to be systematic in the way that we analyse it.

Let us start with $\underline{i}$, $\underline{j}$ and $\underline{k}$ as the usual x, y, z axes fixed in the mounting of the robot and call this 'Frame 0'. We need to know the transformation that will convert the coordinates of anything held in the gripper into coordinates with respect to the reference Frame 0 in the robot's base.

We can define a succession of frames as we make our way along the robot to the gripper. Each of these frames will have a local x, y and z direction related by some transformation to the next frame. Some transformations will relate to the variable angles that make up the axes, others will simply take us from one end to the other of a link such as the 'forearm'.

We can choose the frames so that the transformations between them are extremely simple, being either a rotation about one of the axes or a translation along one of the axes.

Fig. 9.2 Unimation Puma

Let us see it in action.

The joints of the Puma can be thought of as mimicking the human body. The first joint is a 'waist joint' that rotates the whole of the rest of the robot about a vertical axis.

Then, mounted a little to one side is a simplified 'shoulder joint'. This allows the upper arm to rotate about a horizontal axis extending from the 'shoulder'.

Next we have a simplified 'elbow joint', also allowing rotation about a horizontal axis parallel to that of the shoulder.

Then we have three wrist joints to which it is difficult to assign names. The first allows rotation about the line of the forearm, as you would use when turning a door handle. The second is a hinge perpendicular to this, such as you might use when patting a dog. The third is a twist, rather like a screwdriver held between fingers and thumb.

We need to define a chain of frames all the way from Frame 0 to the gripper.

For our first 'journey', let us simply rise up the shaft of the robot to the height of the shoulder, where we will put Frame 1. The transformation $^0T_1$ will convert Frame 1 coordinates to Frame 0 coordinates, and so will be

$$^0T_1 = \begin{bmatrix} 1 & 0 & 0 & 0 \\ 0 & 1 & 0 & 0 \\ 0 & 0 & 1 & h \\ 0 & 0 & 0 & 1 \end{bmatrix}$$

where h is the height of the shoulder from the base. This transformation will simply add h to the z coordinate.

Now we will use the waist joint to rotate the line of the shoulder. This is a rotation about the z axis through an angle θ_1 and we will use our shorthand notation. Frame 2 will now be at shoulder height with the y axis along the line of the shoulder pivot:

$$^1T_2 = \begin{bmatrix} c_1 & -s_1 & 0 & 0 \\ s_1 & c_1 & 0 & 0 \\ 0 & 0 & 1 & 0 \\ 0 & 0 & 0 & 1 \end{bmatrix}$$

Now, since the upper arm is offset from the shoulder, for Frame 3 we should step in the y direction to the line of the upper arm, distance a.

$$^2T_3 = \begin{bmatrix} 1 & 0 & 0 & 0 \\ 0 & 1 & 0 & a \\ 0 & 0 & 1 & 0 \\ 0 & 0 & 0 & 1 \end{bmatrix}$$

Now the shoulder axis rotates the upper arm θ_2 about the y-axis of Frame 3, so we align Frame 4 with that limb. But should we align it with x or z? It seems logical to measure the arm's angles up and down from 'straight out', so we choose x.

$$^3T_4 = \begin{bmatrix} c_2 & 0 & s_2 & 0 \\ 0 & 1 & 0 & 0 \\ -s_2 & 0 & c_2 & 0 \\ 0 & 0 & 0 & 1 \end{bmatrix}$$

Now we must 'move down the upper arm' to the elbow, a distance l, say. This is in the x-direction of Frame 4, so

$$^4T_5 = \begin{bmatrix} 1 & 0 & 0 & l \\ 0 & 1 & 0 & 0 \\ 0 & 0 & 1 & 0 \\ 0 & 0 & 0 & 1 \end{bmatrix}$$

In the Puma, the forearm is offset slightly from the upper arm, but to avoid adding an extra frame we can take account of this in the value of a, above.

So now let us bend the elbow through θ_3 and line up Frame 6 with the forearm. Once again the pivot is the y-axis and zero deflection is taken as 'elbow straight'.

$$^5T_6 = \begin{bmatrix} c_3 & 0 & s_3 & 0 \\ 0 & 1 & 0 & 0 \\ -s_3 & 0 & c_3 & 0 \\ 0 & 0 & 0 & 1 \end{bmatrix}$$

Frame 7 is lined up with the forearm, but has moved down to the wrist, distance m

$$^6T_7 = \begin{bmatrix} 1 & 0 & 0 & m \\ 0 & 1 & 0 & 0 \\ 0 & 0 & 1 & 0 \\ 0 & 0 & 0 & 1 \end{bmatrix}$$

Frame 8 follows the first wrist rotation, θ_4 about the local x-axis.
Frame 9 'waves farewell', θ_5 about the local y-axis.
Frame 10 'twists the screwdriver' θ_6 about the local x-axis.
Frame 11 'reaches' to the end of the screwdriver.
As an exercise, write down the corresponding transformations.
So just what do we do with all these matrices?
Each matrix transforms the coordinates to the 'next lower' frame of reference, the final transformation bringing us to the reference Frame 0. But remember that the matrices are stacked up right to left, the first to be applied being closest to the vector it multiplies, which in this case is the coordinate of a point with respect to the gripper axes. So the product ends up as

$$^0T_{11} = {}^0T_1 \, {}^1T_2 \, {}^2T_3 \, {}^3T_4 \, {}^4T_5 \, {}^5T_6 \, {}^6T_7 \, {}^7T_8 \, {}^8T_9 \, {}^9T_{10} \, {}^{10}T_{11}$$

We have more matrices to multiply than there are axes, but they are all elementary rotations about an axis or translation along an axis. A prismatic joint appears no different from translation along a limb. The only difference is that the distance parameter will be a variable.

Although the final matrix will be unique, there can be many ways to get there. Rotations about the y axis can be changed so that the 'travel' along a limb is in the z direction, rather than x. But when the matrices are all multiplied together they must give the same result.

There is a JavaScript simulation at http://essdyn.com/sim/makearm.htm where you can see the four-by-four transformations in action. In this example the transformations are built up from the base, as you will see by viewing the source. Each new component is multiplied by the transform 'so far', the product of the transformations that went before it. It is then added to the display.

But there is another methodology that aims to involve just one matrix for each actuated axis. The matrices are not primitives, as above, but are generally the product of three elementary moves.

9.4 D-H Parameters

The mechanism consists of a chain of links between one axis and the next. The Denavit Hartenburg convention is based on making all rotations and prismatic actuations take place about the z-axis of a frame.

- We have a set of axes at each joint. The z-axes z_{n-1} and z_n at each end of link n are aligned with the axis of rotation or translation there.
- The x-axis x_n at the 'outer end' is chosen so that it is normal to both of these z axes.
- Now that we know x_n and z_n, we can define y_n to be perpendicular to these to make up a right-handed set of axes.
- If the z axes are not parallel, the transformation for that link must include a 'twist' α about the x-axis.
- The translation will consist not only of a displacement l in the x direction, but can also have a z component d to account for an offset between the points where the 'previous' and the 'next' normals intersect the z-axis.

For a rotation θ about the first of these z axes, this results in a transformation matrix between these frames

$$
\begin{bmatrix} \cos\theta & -\sin\theta & 0 & 0 \\ \sin\theta & \cos\theta & 0 & 0 \\ 0 & 0 & 1 & 0 \\ 0 & 0 & 0 & 1 \end{bmatrix}
\begin{bmatrix} 1 & 0 & 0 & 0 \\ 0 & 1 & 0 & 0 \\ 0 & 0 & 1 & d \\ 0 & 0 & 0 & 1 \end{bmatrix}
\begin{bmatrix} 1 & 0 & 0 & 0 \\ 0 & \cos\alpha & -\sin\alpha & 0 \\ 0 & \sin\alpha & \cos\alpha & 0 \\ 0 & 0 & 0 & 1 \end{bmatrix}
$$

$$
= \begin{bmatrix} \cos\theta & -\sin\theta\cos\alpha & \sin\theta\sin\alpha & l\cos\theta \\ \sin\theta & \cos\theta\cos\alpha & -\cos\theta\sin\alpha & l\sin\theta \\ 0 & \sin\alpha & \cos\alpha & d \\ 0 & 0 & 0 & 1 \end{bmatrix}
$$

The link transformation can thus be defined by a set of D-H parameters: the actuation angle θ, the link length l, the link offset d and the twist α.

But with the slightest change in the convention, the 'formula' for the transformation will be changed.

It is my opinion that the approach of chaining a set of elementary transformations is safer and better.

9.5 Example of a Simple Kinematic Chain

Figure 9.3 shows a robot arm that can reach points in three dimensions. A coordinate system is defined such that z is vertically upwards, x is forwards and y is to the left.

The first axis, rotation A about the z-axis, can be thought of as a shoulder joint, where the upper arm is kept horizontal. When the angle A is zero, this part of the arm is in the x direction.

The third axis is an elbow joint. When angle C is zero, the arm is straight.

The second axis is a rotation of the upper arm about its length. When angle B is zero, a positive rotation of elbow joint C will at first lower the 'hand' vertically.

The upper arm is of length 0.5 m while the forearm is of length 0.3 m.

(a) **Assuming that there are no limits on the angles, describe the space that can be reached by the 'hand'.**

   If we consider just the elbow and the 'arm twist', the hand can reach any point of a sphere of radius 0.3, centred on the elbow. Now when we also consider the first axis, rotating about z, this sphere will sweep out a torus. The 'hole in the doughnut' will be the difference in the arm lengths, 0.2 m, while the extreme outside will be their sum, 0.8 m. The vertical thickness will be 0.6 m.

(b) **Derive with clear explanation a chain of individual elementary transformations for each rotation and translation.**

   A description of the operations is

   (rotate A about z)(rotate upper arm B about x)
   (translate 0.5 along x to elbow)(elbow angle, rotate C about y)
   (translate 0.3 along x to hand).

Fig. 9.3 An arm to try

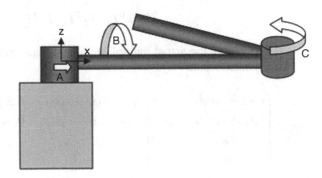

In matrix terms, with c_A for cos(A) etc., these will be

$$^0T_1 = \begin{bmatrix} c_A & -s_A & 0 & 0 \\ s_A & c_A & 0 & 0 \\ 0 & 0 & 1 & 0 \\ 0 & 0 & 0 & 1 \end{bmatrix}$$

$$^1T_2 = \begin{bmatrix} 1 & 0 & 0 & 0 \\ 0 & c_B & -s_B & 0 \\ 0 & s_B & c_B & 0 \\ 0 & 0 & 0 & 1 \end{bmatrix}$$

$$^2T_3 = \begin{bmatrix} 1 & 0 & 0 & 0.5 \\ 0 & 1 & 0 & 0 \\ 0 & 0 & 1 & 0 \\ 0 & 0 & 0 & 1 \end{bmatrix}$$

$$^3T_4 = \begin{bmatrix} c_C & 0 & s_C & 0 \\ 0 & 1 & 0 & 0 \\ -s_C & 0 & c_C & 0 \\ 0 & 0 & 0 & 1 \end{bmatrix}$$

$$^4T_5 = \begin{bmatrix} 1 & 0 & 0 & 0.3 \\ 0 & 1 & 0 & 0 \\ 0 & 0 & 1 & 0 \\ 0 & 0 & 0 & 1 \end{bmatrix}$$

(c) **Multiply these transformations in the correct order, to obtain the transformation matrix that expresses the location and orientation of hand coordinates in terms of the x, y and z axes at the 'shoulder', as indicated.** The correct order is of course

$$^0T_5 = {}^0T_1 \, {}^1T_2 \, {}^2T_3 \, {}^3T_4 \, {}^4T_5$$

Performing the actual multiplication is an arduous task that will produce a complicated result.

$$^0T_5 = \begin{bmatrix} c_A c_C - s_A s_B s_C & -s_A c_B & c_A s_C + s_A s_B c_C & 0.5 c_A + 0.3(c_A c_C - s_A s_B s_C) \\ s_A c_C + c_A s_B s_C & s_A c_B + c_A s_B & s_A s_C - c_A s_B c_C & 0.5 s_A + 0.3(s_A c_C + c_A s_B s_C) \\ -c_B s_C & 0 & c_B c_C & -0.3 c_B s_C \\ 0 & 0 & 0 & 1 \end{bmatrix}$$

I have very likely made a slip. I will leave it to you to check it.

Solving the final part is more easily performed with brain-power than algebra.

(d) **How would you find the joint angles to reach a point $(x, y, z, 1)'$?**

Fortunately in this case there is a simpler solution than struggling with the algebra.

Since you know the coordinates of the target, you know its distance from the origin. The cosine law tells you the square of this distance is

$$d^2 = 0.5^2 + 0.3^2 - 2 \times 0.5 \times 0.3 \, \cos(\pi - C)$$
$$d^2 = 0.34 + 0.3 \, \cos(C)$$

so

$$\cos(C) = (x^2 + y^2 + z^2 - 0.34)/0.3$$

But of course this has two solutions. Suppose that z is negative, then the elbow can either bend down to the target, or the forearm can rotate through 180° and the elbow can bend to a reflex angle. So we choose one of the options for C.

Now since we know C we can find B from

$$y = 0.3 \, \sin(C)\cos(B)$$

And again we have a choice.

But now the horizontal 'reach' r of the arm is known, $r = 0.5 + 0.3 \, \cos(C)$ and we know x and y, so we have just one direction for the 'reach vector'.

Angle A will be this vector direction, $\tan^{-1}(y/x)$, offset by the angle between the reach and the upper arm, $\sin^{-1}(y/r)$. This time the choice has already been made by our choice for the arm rotation.

9.6 Inverse Kinematics

Of course, calculating the kinematics of the robot is only half the story. We can now express the location and orientation of the gripper in terms of the axis movements, but what we really want is to find the axis values needed to put the gripper in some desired position. This calculation is referred to as 'inverse kinematics'.

To find the required joint angles, we can calculate the transformation representing the desired position and then compare coefficients with the general transformation that is full of sines and cosines of those joint angles. That leaves us with some unpleasant simultaneous equations to solve. In fact the result of aligning the three rotations of the wrist joint of the Puma through the same point reduces the algebra and trigonometry significantly. Nevertheless the solutions are not unique.

For any given gripper position and attitude, there is an 'elbow up' solution as well as an 'elbow down' one. These are doubled again with 'lefty' and 'righty'. By 'turning its back' on the work, the robot can turn its single 'right arm' into a left one.

Then, of course, not all positions have a solution. The desired position might be just out of reach of the outstretched arm.

Another problem is 'singularity'. The robot normally has six degrees of freedom. But when two joints are in line, such as the wrist and 'screwdriver twist', the degrees of freedom drop to five. In the neighbourhood of a singularity, one of the axes will have to move rapidly for the slightest change of the target position.

Think of the problem of trying to watch an aircraft as it flies past straight overhead. Your head moves up until the plane is overhead, then you must spin through 180° to watch it fly away.

Of course the robot might not have six axes and we might not wish to move in all six degrees of freedom.

For example a 'pick and place' robot might be concerned with placing components on a circuit board. The components are presented 'flat', so we have no need to tilt them. We might, however, need to rotate them about a vertical axis to align them with the board and we need to move them to an accurate x-y position. We need a fourth axis to lift them above the board before we place them, but this might just travel between two stops.

Clearly, for a solution to 'make sense', there must be the same number of control axes as we wish to obtain degrees of freedom. But what if our robot has seven axes?

For various reasons, extra axes might be added, perhaps to allow the robot to 'reach around corners'. In this case a unique solution is impossible, not even giving a choice of one in four. To extract a solution, an extra condition has to be imposed, such as that one axis is held at an extreme or at zero deflection.

But an alternative method is outlined in the next chapter.

9.7 A SCARA Example

Example 9.1
(a) **A surface-mountable chip is to be placed on a circuit board. What are the degrees of freedom that its final position will have?**
Figure 9.4 shows a 'SCARA' robot to be used for automatic placement. There are rotary joints with vertical axes at A, B and C and a telescopic vertical joint D. The horizontal rods are of length L and M.
(b) **Derive with clear explanation a chain of individual elementary transformations for each rotation, change of coordinates along a rod or actuation by joint D.**

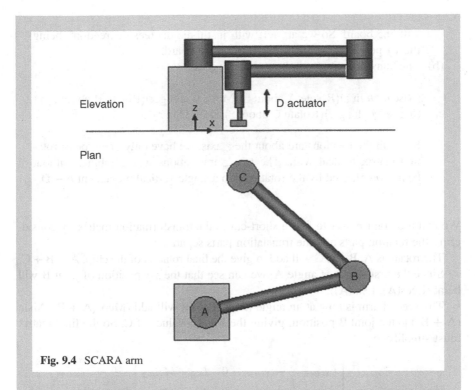

Fig. 9.4 SCARA arm

Use the abbreviations s1, c1, s2, c2, s3 and c3 for the sines and cosines of the angles of rotation of the axes at A, B and C, where joints A, B and C will lie in a straight line to the right of A when all angles are zero.

(The effective height of the top of actuator D is h, so that the vertical position of the chip is $h - D$.)

(c) **Multiply these transformations in the correct order, to obtain the transformation matrix that expresses the location and orientation of the chip in terms of the x, y and z axes at the base of the pillar, as indicated. Simplify the expressions.**

An incidental comment: The acronym 'SCARA' stands for 'Synthetic Compliance Adaprice Robot Aem'. But few if any of the robots with a SCARA configuration possess this attribute! The label has been attached to any robot that has only vertical revolute joints, like the pick-and-place robot in question.

Solution

(a) Of the original six degrees of freedom of an unrestrained body, the height will be constrained when the chip is attached, so one is lost, while two degrees of rotation are lost because the chip must be parallel with the board. So we are left with just three degrees of freedom, being the x y position and the orientation on the board.

(b) The transformations can be summed up as

(Rise by h in z)(Rotate A about z)(Move L along x)(Rotate B about z)
(Move M along x)(Rotate C about z)(Drop D in z)

(c) Since all the rotations are about the z axis, we have only one type of rotation matrix to deal with. The two z translations will combine, without being complicated by the rotations, to a single vertical movement $h - D$.

We can use brain-power to find a short-cut to the transformation matrix by considering the rotation parts and the translation parts separately.

The rotations A, B and C will add to give the final rotation of the chip, $A + B + C$.

Since the first arm is at angle A, we can see that the x-y position of joint B will be at $(Lcos(A), Lsin(A))$.

The second arm is now at an angle $(A + B)$ so it will add $(Mcos(A + B), Msin(A + B))$ to the joint B position, giving the x and y values of C. So the final matrix must simplify to

$$
\begin{bmatrix}
c_{A+B+C} & -s_{A+B+C} & 0 & Lc_A + Mc_{A+B} \\
s_{A+B+C} & c_{A+B+C} & 0 & Ls_A + Ms_{A+B} \\
0 & 0 & 1 & h - D \\
0 & 0 & 0 & 1
\end{bmatrix}
$$

Chapter 10
Inverse Kinematics

Abstract This short chapter introduces you to some techniques and concepts that you might find useful. Rather than trying to solve for the axis angles in a single hit, we 'creep up on them' by asking whether a positive or a negative 'twitch' will get us closer to the target. The Jacobian gives a more formal way to express this.

10.1 Introduction

This short section looks at the way partial derivatives can reveal the response of the robot to axis deflections, in terms of velocities and incremental deflections.

10.2 The Inverse Kinematics Problem

From the kinematics, we have found a chain of matrices that can be multiplied together to obtain the transformation matrix describing the motion of a robot. The right hand column defines the location of the origin of the gripper, $(x, y, z)'$, while a three-by-three sub-matrix tells us the gripper's orientation in terms of the unit vectors $\underline{n}$, $\underline{o}$ and $\underline{a}$.

$$\begin{bmatrix} n_x & o_x & a_x & x \\ n_y & o_y & a_y & y \\ n_z & o_z & a_z & z \\ 0 & 0 & 0 & 1 \end{bmatrix}$$

But this dozen coefficients is the result of rotating the gripper through the three angles of pitch, roll and yaw and translating it to coordinates of x, y and z. So we

© Springer International Publishing AG 2018
J. Billingsley, *Essentials of Dynamics and Vibrations*,
DOI 10.1007/978-3-319-56517-0_10

should be able to define our transformation in terms of a vector with just six components:

$$(x, y, z, \theta, \varphi, \psi)'$$

Each of these components will be a function of all six joint axes

$$x(\theta_1, \theta_2, \theta_3, \theta_4, \theta_5, \theta_6)$$
$$y(\theta_1, \theta_2, \theta_3, \theta_4, \theta_5, \theta_6)$$
$$z(\theta_1, \theta_2, \theta_3, \theta_4, \theta_5, \theta_6)$$
$$\theta(\theta_1, \theta_2, \theta_3, \theta_4, \theta_5, \theta_6)$$
$$\varphi(\theta_1, \theta_2, \theta_3, \theta_4, \theta_5, \theta_6)$$
$$\psi(\theta_1, \theta_2, \theta_3, \theta_4, \theta_5, \theta_6)$$

These tell us the 'kinematics' of the system, the end-effector position and orientation that will result from a specific set of joint angles. However in designing a robot system, we know very well where we want the end effector to be but we want to know the joint angles that will give us that position. This is the Inverse Kinematics problem.

As you were able to see from the equations that describe the Puma, unscrambling the equations is not easy. Matters are made worse if there are more than six control axes, or if the dynamics have become singular. But there are other techniques than inverting the algebra.

10.3 Making Use of the Jacobian

Although solving equations to find functions for the axis values might not be easy, we can find the effect of a 'twitch' in one of the axes by partial differentiation.

If we change just θ_1 by $\delta\theta_1$, the change in x will be

$$\delta x = \frac{\partial x}{\partial \theta_1} \delta\theta_1$$

If we change more of the joints, we will have

$$\delta x = \frac{\partial x}{\partial \theta_1} \delta\theta_1 + \frac{\partial x}{\partial \theta_2} \delta\theta_2 + \frac{\partial x}{\partial \theta_3} \delta\theta_3 + \frac{\partial x}{\partial \theta_4} \delta\theta_4 + \frac{\partial x}{\partial \theta_5} \delta\theta_5 + \frac{\partial x}{\partial \theta_6} \delta\theta_6$$

In fact we can calculate all the partial derivatives and find the Jacobian, a matrix that has all the partial derivatives as its coefficients

$$
J = \begin{bmatrix}
\dfrac{\partial x}{\partial \theta_1} & \dfrac{\partial x}{\partial \theta_2} & \dfrac{\partial x}{\partial \theta_3} & \cdots & \cdots & \cdots \\[2ex]
\dfrac{\partial y}{\partial \theta_1} & \dfrac{\partial y}{\partial \theta_2} & \dfrac{\partial y}{\partial \theta_3} & \cdots & \cdots & \\[2ex]
\dfrac{\partial z}{\partial \theta_1} & \dfrac{\partial z}{\partial \theta_2} & \dfrac{\partial z}{\partial \theta_3} & \cdots & & \\[2ex]
\dfrac{\partial \theta}{\partial \theta_1} & \dfrac{\partial \theta}{\partial \theta_2} & \dfrac{\partial \theta}{\partial \theta_3} & \cdots & & \\[2ex]
\dfrac{\partial \phi}{\partial \theta_1} & \dfrac{\partial \phi}{\partial \theta_2} & \dfrac{\partial \phi}{\partial \theta_3} & \cdots & & \\[2ex]
\dfrac{\partial \psi}{\partial \theta_1} & \dfrac{\partial \psi}{\partial \theta_2} & \dfrac{\partial \psi}{\partial \theta_3} & \cdots & &
\end{bmatrix}
$$

Now at any given position, these coefficients will just be numbers that we can calculate, so that we can find the effect of 'nudging' the joints from the equation

$$
\begin{bmatrix}
\delta x \\
\delta y \\
\delta z \\
\delta \theta \\
\delta \phi \\
\delta \psi
\end{bmatrix}
= J
\begin{bmatrix}
\delta \theta_1 \\
\delta \theta_2 \\
\delta \theta_3 \\
\delta \theta_4 \\
\delta \theta_5 \\
\delta \theta_6
\end{bmatrix}
$$

If we can invert this numerical Jacobian, we can find the adjustments in the joint angles to bring us closer to the target

$$
\begin{bmatrix}
\delta \theta_1 \\
\delta \theta_2 \\
\delta \theta_3 \\
\delta \theta_4 \\
\delta \theta_5 \\
\delta \theta_6
\end{bmatrix}
= J^{-1}
\begin{bmatrix}
\delta x \\
\delta y \\
\delta z \\
\delta \theta \\
\delta \phi \\
\delta \psi
\end{bmatrix}
$$

We will probably not get to the target exactly because the equations are non-linear, but successive approximations will get us as close as we need.

If we have planned a path for a tool piece, we will have a way to compute a succession of 'way points', so the distance from one target to the next will be small. If we are off the target, we know the values we need to approach it, at least to a first approximation.

Often this will work! But it is possible that the Jacobian is singular, meaning that it has no finite inverse. This will happen at a singularity.

All is not lost. A method of successive approximations can bring us closer to the target, or to the point in the 'reachable' space that is closest to it. For each axis in turn, inspect the corresponding column of the Jacobian and decide whether a positive or a negative nudge will bring us closer to the target, or whether that axis should remain the same. Apply the nudges and measure the new error. When there is no sign of improvement, halve the nudge size.

The Jacobian also relates the gripper velocity to the velocities of the axes. If the objective is to move the gripper along a path at maximum speed, one or more of the axes will be required to reach maximum velocity. As the gripper moves along the path, the identity of the limiting axis will probably change. Once again, the Jacobian and its inverse will be valuable tools in calculating the axis drive values.

10.4 Singularity

As each axis is 'twitched' there is a resulting vector movement of the end-effector – in all six parameters. With six axes and six degrees of freedom, all should be well. But if two of the axes make the end-effector change in exactly the same direction, the system has become singular. This happens, for instance, when the first wrist joint and the 'screwdriver twiddle' joint are lined up on a Puma.

It also happens when the elbow is absolutely straight, so that the result of elbow movement is in exactly the same direction as for shoulder movement.

Singularity will also usually be seen at the 'limit of reach', the boundary of the region the robot can access.

Another form of singularity occurs when the robot has too many axes! Now we can find solutions to put the gripper where we want, but there are an infinite number of solutions unless we impose an arbitrary constraint. The Jacobian is singular, being perhaps of dimensions 6 by 7, so we cannot invert it. But with some ingenuity we can manage to find an answer.

Chapter 11
Vibration

Abstract This chapter is all about vibration.

- When there is a force or couple that accelerates something towards some central position, things vibrate.
- If there is no 'damping' to soak up energy, they will vibrate forever.
- If the force or couple is proportional to the displacement, then the movement will be a sinusoidal (sine or cosine) function of time.
- If there is some damping, with a force or couple proportional to velocity, the sine-wave will 'decay' exponentially.

As an engineer, you will probably wish to add some damping to limit any unwanted vibration or to change the 'resonant' frequency away from that of a disturbance.

As a musician, you would wish to shape the vibration to something that sounds pleasant!

So to deal with a vibrating system you should:

1. **Look for the 'variables' that describe what is happening**.
2. **Find some equations for their rates-of-change in terms of all such variables and any inputs**.
3. **Either eliminate all the variables but one, to get a differential equation, or else find the characteristic equation of a first order matrix equation to find the eigenvalues**.
4. **Solve this equation to analyse what will happen, in terms of the frequencies involved**.

Now there are two ways to go about step 3. You can mess about with simultaneous equations and algebra, or you can use the power of matrices to help you.

- Write the 'state equations' in matrix form.
- Look for 'eigenvalues' and 'eigenvectors', where the rate-of-change of an eigenvector is just the eigenvalue times the vector itself.

This gives an exponential solution that is easy to recognise.

© Springer International Publishing AG 2018
J. Billingsley, *Essentials of Dynamics and Vibrations*,
DOI 10.1007/978-3-319-56517-0_11

11.1 Introduction

Oscillating systems are everywhere! Wherever there is an acceleration opposing a deflection, being proportional to that deflection, you will find 'simple harmonic motion'.

The most obvious example is the pendulum, where if the angle x is small, we can write

$$x'' = -gx/L$$

where the double dash (or double dot) indicates a second derivative with respect to time.

The writers of books on dynamics are very fond of springs and masses, giving very similar equations. Of course bells, stringed instruments and shopping-trolley casters can join the collection.

Except in musical instruments and clocks, oscillations are usually regarded as undesirable. They will usually die away because of losses in the system, but engineers will often wish to make them die away faster, or even damp out the oscillations completely.

11.2 Vibrating Systems

Fig. 11.1 Vertical oscillation

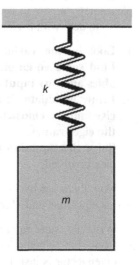

Consider a mass that is hanging on a spring. Let us forget about all the degrees of freedom except one, vertical motion. If the mass is m and the spring stiffness is k, then the mass can be at rest when the spring has been extended by mg/k. So let us measure any small deflection from that rest position as x.

If the mass is raised by distance x, the spring tension will reduce by kx, so that the mass will accelerate downwards at a rate

$$x'' = -kx/m$$

But the example that you are more likely to find in textbooks involves a trolley.

Fig. 11.2 Horizontal oscillation

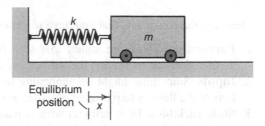

We will get the same equation for this frictionless trolley of mass m, rolling on a horizontal track and connected to some fixed point by a spring of stiffness k.

The advantage of this example is that we can consider adding a 'damper'.

Fig. 11.3 Horizontal oscillation with damping

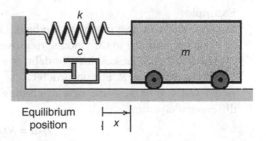

A damper is rather like a hydraulic cylinder with an internal leak. When a force is applied to it, it will change length at a velocity proportional to the force. We can say that the force is proportional to the rate-of-change of length according to

$$f = -c\,x'$$

so that the equation for the trolley with damper becomes

$$x'' = -\frac{k}{m}x - \frac{c}{m}x'$$

To investigate this kind of system we need to be able to deal with equations of the form

$$x'' + a\,x' + b\,x = 0$$

or if we apply some disturbance to it,

$$x'' + a\,x' + b\,x = f(t)$$

Details of solving differential equations are given in the maths revision material, but perhaps we should outline them here. In particular we will be interested in the role of exponentials in the solution.

11.3 Differential Equations and Exponentials

There are three concepts that you have to get clear in your mind:

1. **Parameters**. These are values that stay fixed as time goes on, such as the acceleration due to gravity, the stiffness of a spring or the length of a string.
2. **Inputs**. Something might be happening to disturb you system, like the vibration of the floor, a step change in target or some vibrating or random force.
3. **State variables**. These tell you what is really going on. As a pendulum swings its angle changes, as a mass bounces its position changes too, but you usually also have to know some rates-of-change, because velocities cannot change instantaneously.

Examples

Suppose that you owe money to the bank. The bank will charge you interest at a rate proportional to the amount you owe, multiplied by their interest rate.

The owed amount, call it D for debt, is a **state variable**.

The interest rate, R, is a **parameter**.

So you can express the **rate-of-change** of the amount you owe by the differential equation

$$dD/dt = RD$$

And if you spend some more money, S, that is an input, but in this case it relates to an 'event' and does not fit the idea of a rate-of-change.

So 'real' problems will start with the task of spotting **state variables** and the **state equations** that define their rate of change.

See if you can spot the state variables and state equations in these examples.

1. A cooling cup of coffee.
2. An object falling vertically under gravity.

The bank account example is a 'first order system'. There is just one state variable and one state equation. The solution will involve an exponential $D_0 e^{Rt}$, where D_0 is the amount you owed to start off with. If you leave your debt unpaid for a long time it will grow to be huge, but as differential equations go this is not really interesting.

The cup of coffee is also a first-order system with one state variable, the excess temperature. The difference is that in this case the coffee cools toward a limit, rather than growing exponentially away from one. The exponential in the solution will be $T_0 e^{-t/\tau}$, where T represents the difference from the ambient temperature and τ is the time it takes for the excess temperature to reduce by a factor of e.

But to examine vibrating systems we must look at second order systems. We will be concerned with velocities, but also we will involve accelerations.

Hold a ball at arm's length, then let it go. If we just consider its vertical movement we have two state variables, its height x and its vertical velocity, v. Its vertical acceleration is $-g$, a fixed parameter.

If your mind is set on second-order equations you might write

$$x'' = -g$$

where we use a double dash or a double dot to denote the second derivative with respect to time. It is easier to type than d^2x/dt^2.

But with state-variables in mind we can write it as two first-order equations

$$x' = v$$
$$v' = -g$$

What actually happens will depend on the initial conditions. If we throw the ball upwards, v will start with a positive value. We can solve it simply by integrating twice to get

$$x = x_0 + v_0 t - g t^2 / 2.$$

We can simulate the system with a couple of lines of code, such as

```
x = x + v*dt;
v = v - g*dt;
```

which are executed over and over again to advance time in small steps of dt. If we plot x against t we will see a parabola.

Do not be afraid of the snippets of code that keep cropping up – you will not be asked to write any JavaScript!

But this is an easy way to put dynamic examples on the page to show you how things move – and you will see that you can easily make simulations of your own.

The essential concept is just that a value of something a moment dt later will just be its value now plus dt times its rate-of-change.

11.4 Simple Harmonic Motion

Things get more interesting when the acceleration depends on the displacement x itself

$$x'' = -n^2 x.$$

This is the expression we have met in school maths or physics concerning simple harmonic motion, SHM. It can apply to a pendulum of length L, where the acceleration back towards the centre is proportional to the displacement and where $n^2 = g/L$.

Simple Harmonic Motion can also apply to the systems with a mass and a spring that are illustrated in Figs. 11.1, 11.2 and 11.3.

If the stiffness of the spring is k N/m, it means that if the spring is compressed by an amount x it will exert a force kx N.

If this force is applied to a mass m, it will result in an acceleration of $-kx/m$, since the spring force opposes the displacement.

So once more we have

$$x'' = -n^2 x,$$

where in this case $n^2 = k/m$.

One solution would be $x = \cos(nt)$.

The first derivative is $-n \times \sin(nt)$ and the second is $-n^2\cos(nt)$ – just what we need, being equal to $-n^2 x$.

It could equally be $x = \sin(nt)$, with first derivative $n \cos(nt)$ and second derivative $-n^2\sin(nt)$.

In general it is a mixture of the two, $A.\cos(nt) + B.\sin(nt)$, where A and B are determined by the initial conditions.

You will usually find the symbol ω (omega) used for the frequency instead of n.

Note that omega is an 'angular frequency' in rad/s. To get Hertz (cycles per second) you have to divide by 2π.

The solution gets more complicated when we include 'damping', where the acceleration is determined by a combination of both position and velocity, of the form

$$x'' = -n^2 x - av$$

In state-equation terms, the complication is minimal.

We simply write the equations as

$$x' = v$$
$$v' = -n^2 x - av$$

where v is the velocity and a is a constant that multiplies it to get the damping force.

This would represent the second-order equation

$$x'' + ax' + n^2 x = 0$$

Very often we will find that instead of a zero on the right hand side, there is some function of time that is a **forcing function** input

$$x'' + ax' + n^2 x = f(t)$$

11.5   More on Exponentials

In general, if we have $x' = kx$ we will look for an e^{st} solution and find that $s = k$.
Each time we differentiate we pick up another s, so for the pendulum where

$$x'' = -(g/L)x$$

we would have

$$s^2 e^{st} = -(g/L)e^{st}$$

i.e.

$$s = \pm\sqrt{(-g/L)}$$

But that is the root of a negative quantity, so s is an imaginary number, $\pm j\omega$.
Don't panic!
It can be shown that

$$e^{j\omega t} = \cos(\omega t) + j\sin(\omega t)$$

and

$$e^{-j\omega t} = \cos(\omega t) - j\sin(\omega t)$$

so this is just leading us back to the sines and cosines that we already knew would
form the solution.

But it's easier than that.

By taking multiples of these, where the multipliers are themselves complex
numbers, we can get any mixture of sines and cosines that the initial conditions
will dictate.

If we just take

$$(a + jb)e^{j\omega t}$$

we will get something that has a real part equal to

$$a\cos(\omega t) - b\sin(\omega t)$$

which represents any mixture that we like. Instead of solving simultaneous
equations with sines and cosines we just need to look for a single complex number
to multiply $e^{-j\omega t}$, so that

$$x = \Re(Ce^{j\omega t})$$

where C is complex and the 'funny R' means 'the real part of'.

11.6 Adding Damping

Although the damper in a car's tailgate lift is air filled, the most common damper
for systems like the trolley uses a fluid filled cylinder. This has a piston with holes
or other oil paths. When the length changes, oil is driven through the holes, losing
energy in the process. The rate-of-change of length is proportional to the applied
force, so we can define a parameter c such that

$$F = -c \, \mathrm{d}(\text{length})/\mathrm{d}t$$

where c will have the units Newtons per (metre per second), written as Ns/m.

Now when we look at systems with damping, things get more interesting, but
not really more difficult.

Fig. 11.4 Trolley with
damping

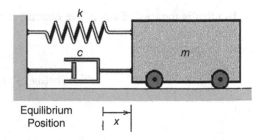

In the 'damped mass spring' example we found that

$$x'' + (c/m)x' + (k/m)x = 0$$

When we try $x = e^{st}$ we find that

$$s^2 + (c/m)s + k/m = 0$$

so using the formula for solving a quadratic we have

$$s = -c/2m \pm \sqrt{(-k/m + c^2/4m^2)}$$

If c/m is 'not too big' the expression in the square root will be negative, so we
will have an answer that is not just imaginary, it is complex.

$$s = \sigma + j\omega$$

where

$$\sigma = -c/2m$$

and

$$\omega = \sqrt{(k/m - c^2/4m^2)}$$

Now $e^{(\sigma+j\omega)t}$ can be split into the product of two parts

$$e^{\sigma t} \text{ times } e^{j\omega t}.$$

It means that the sine-wave that we saw before is now multiplied by an exponential which will usually represent something dying away with time.

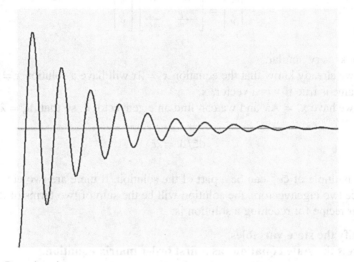

Fig. 11.5 Decaying sinewave

Try to keep these points in mind. They will be repeated in the material that follows.

11.7 Matrices Can Help

Instead of reducing the equations for the trolley and spring to a second order equation we can write them as two first-order equations

$$\dot{x} = v$$

$$\dot{v} = -\frac{k}{m}x$$

where the 'dot' represents d/dt.

We can write these two equations in matrix form as

$$\begin{bmatrix} \dot{x} \\ \dot{v} \end{bmatrix} = \begin{bmatrix} 0 & 1 \\ -\frac{k}{m} & 0 \end{bmatrix} \begin{bmatrix} x \\ v \end{bmatrix}$$

This makes things much easier when there is damping, or when we want to simulate anything.

For the example where a damper has been added to the trolley and spring, we get

$$\begin{bmatrix} \dot{x} \\ \dot{v} \end{bmatrix} = \begin{bmatrix} 0 & 1 \\ -\frac{k}{m} & -\frac{c}{m} \end{bmatrix} \begin{bmatrix} x \\ v \end{bmatrix}$$

which looks very similar.

Now we already know that the equation $x' = \lambda x$ will have a solution $x = x(0)e^{\lambda t}$. The same is true if x is a vector, $\mathbf{x}$.

So if we have $\mathbf{x}' = \mathbf{A}\mathbf{x}$ and we can find an eigenvector ξ, so that $A\xi = \lambda\xi$, then

$$d\xi/dt = \lambda\xi$$

and so a multiple of $\xi e^{\lambda t}$ can be a part of the solution. If there are two eigenvalues, and hence two eigenvectors, the solution will be the sum of two terms of this sort.

So our recipe for reaching a solution is:

1. **Identify the state variables.**
2. **Express the state equations as a first order matrix equation.**
3. **Find the 'characteristic equation' for that matrix.**
4. **Solve the equation to get the eigenvalues.**
5. **If you want actual solutions, not just frequencies, find the eigenvectors.**
6. **Find values for the a's in the solution, $\Sigma a\xi e^{\lambda t}$ summed over all the eigenvalues, in order to fit it to the initial conditions.**

To make sense of this, let us consider the undamped example.

First we need to calculate the characteristic equation for

$$\begin{bmatrix} 0 & 1 \\ -\frac{k}{m} & 0 \end{bmatrix}$$

To do that, we work out the determinant

$$\begin{vmatrix} 0 - \lambda & 1 \\ -\frac{k}{m} & 0 - \lambda \end{vmatrix} = 0$$

giving

$$\lambda^2 + k/m = 0$$

so

$$\lambda = \pm j\sqrt{(k/m)}$$

This gives us a mixture of sines and cosines of ω, where

$$\omega = \sqrt{(k/m)}.$$

Nothing surprising there!

So let us look at the damped case. Now we will be looking for the eigenvalues of

$$\begin{bmatrix} 0 & 1 \\ -\frac{k}{m} & -\frac{c}{m} \end{bmatrix}$$

and the determinant

$$\begin{vmatrix} 0-\lambda & 1 \\ -\frac{k}{m} & -\frac{c}{m}-\lambda \end{vmatrix}$$

will give us the equation

$$\lambda^2 + (c/m)\lambda + k/m = 0$$

so

$$\lambda = -(c/m)/2 \pm \sqrt{(c^2/4m^2 - k/m)}$$

which we will instead write as

$$\lambda = -(c/m)/2 \pm j\sqrt{(k/m - c^2/4m^2)}$$

Now the expression in the square root might look rather strange when put this way around, but the assumption is that the damping is small enough to give a decaying oscillation. If the damping is 'critical' or above, you will instead get two real roots and the formula will look normal.

Things to take note of:

If we have an equation

$$\lambda^2 + 2a\lambda + b = 0$$

and a is small enough that the roots are complex, then the solution will represent a sine-wave decaying according to e^{-at}. The 'time constant' of the decay is $1/a$.

As the damping a is increased, the frequency part of the solution will decrease. The time-constant will get shorter so that the system settles more quickly.

When a reaches the value $\sqrt{b}$, there will be two equal real roots of $-\sqrt{b}$. The system is said to be 'critically damped'. There is no sinusoidal term.

If a is increased yet further, the system will take longer to settle. We know that the product of the roots must be b, so if one root gets faster then the other must get slower.

11.8   Mathematical Notation

Control engineers have their own notation for this second order equation. They write it as

$$\lambda^2 + 2\zeta\omega_0\lambda + \omega_0^2 = 0$$

corresponding to the differential equation

$$x'' + 2\zeta\omega_0 x' + \omega_0^2 x = 0$$

ω_0 is called the 'undamped natural frequency' for obvious reasons. It is the resonant frequency if there is no damping.

ζ (zeta) is called the 'damping factor'. If it is 0, there is no damping. If it is less than 1, the system is said to be 'underdamped', if it is equal to 1 the system is 'critically damped'. Meanwhile the actual frequency will be $\omega_0\sqrt{(1 - \zeta^2)}$ until ζ reaches unity.

If ζ is greater than 1, there will be two real roots and the settling will get slower as ζ increases.

If ζ stays the same, but ω_0 changes, the time-scale will change, stretching or shrinking the waveform, but the intrinsic shape will remain the same. That is of advantage to the control engineers when they are 'identifying' a system to find out its parameters. From the ratio of one peak to the next, you can calculate a value for ζ, whatever the natural frequency.

You can see plots of the response for various values of zeta at www.essdyn. com/sim/zeta.htm. You will see that the response for $\zeta = 0.8$ appears to settle faster than for critical damping.

For problems such as the ones that you are likely to encounter, where you are given the parameters, the use of this notation is likely to lead to confusion and to disguise the effects that the plain numbers can show.

11.9   Some Examples

Example 11.1 Undamped vibration
The figure below shows a plate with mass $m = 100$ kg. Each of the springs has stiffness $k = 200$ kN/m.

1. By how much will the springs be stretched when the plate is at rest?
2. What will be the frequency (in Hz) for small vertical vibrations?
3. The plate is disturbed from rest by an impulse that gives it a downward velocity of 0.4 m/s. Will the springs go slack at the top of the bounce? (Assume that the mass does not sway or 'twist'.)

Fig. 11.6 Mass on springs

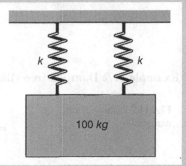

Solution
1. The weight is $100\,g$ N. The springs combine to have stiffness 400 kN/m, so if we approximate g to 10 the deflection will be $100 \times 10/400{,}000 = 2.5$ mm.
2. We have $100\,x'' = -4*10^5\,x$,
   so

$$x'' + 4000\,x = 0$$

$$\omega^2 = 4000$$

$$\omega = 20\sqrt{10}$$

so

$$f = \omega/2\pi = 10 \text{ Hz (approximately)}$$

3. Now since the motion starts from zero deflection we can say that

$$x = a \sin \omega t$$

and

$$x' = a\omega \cos \omega t$$

so since the velocity after the impulse is −0.4 m/s, amplitude $a = -0.4/\omega$.

$$a = -0.4/63.2 = 6.3 \text{ mm}$$

The springs will certainly go slack.

Example 11.2 Damped free vibration

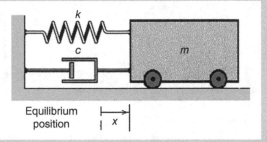

Fig. 11.7 Mass-spring-damper

A classic spring-mass-damper system lying in the horizontal plane has parameters $m = 1$ kg, $k = 25$ N/m, $c = 6$ Ns/m.

The mass is displaced 30 mm from its equilibrium position and released from rest.

1. Is the system under, over or critically damped?
2. Determine the period of motion if it exists.
3. Determine the displacement at $t = 1$ s.

Solution
The differential equation for the system is

$$x'' + 6x' + 25x = 0$$

so we will be looking for the roots of

$$\lambda^2 + 6\lambda + 25 = 0$$

$$\lambda = -3 \pm \sqrt{(9 - 25)}.$$

The solution is $e^{-3t}(A.\cos 4t + B.\sin 4t)$
So the answer to question 1 is 'underdamped'.
The answer to question 2 is period = $2\pi/4 = \pi/2$ s.
For question 3 we must solve for the initial conditions At $t = 0$:

$A = 0.03$ (initial displacement)
$4B - 3A = 0$ (initial velocity)
so $B = 0.0225$

So after 1 second the displacement is

$$0.03\, e^{-3} \cos(4) - 0.0225\, e^{-3} \sin(4)$$

But we can also attack the problem by simulation.

11.10 Simulation Simplified

There is a very simple way to simulate a system using JavaScript (its code looks very similar to C) and you can run it in any browser. You can see a number of examples at www.essdyn.com (though some of them are rather old now).

If you are unfamiliar with C, here are a few pointers:

Every 'assignment', where the value of a variable is changed, must be terminated with a semi-colon.

The 'state variables' (the things that change with time, telling us exactly what is going on) are the position and velocity of the mass, x and v.

The 'state equations' tell us their rates of change, so that we can work out their next value. In this case the equations are

$$dx/dt = v;$$

and

$$dv/dt = force/m;$$

where force is an expression that will involve the present x and v and any input function.

We can update the position by adding (velocity * timestep) to it. This approximation will be more than adequate, provided that the time step is small enough. In a language like Basic, we could write

$$x = x + v * \mathrm{d}t$$

But in C we must add a semicolon:

$$x = x + v * \mathrm{d}t;$$

In addition, C has the alternative 'clever' notation of writing this as

$$x+ = v * \mathrm{d}t;$$

'Jollies' gives us two ways to display the output. The first is by moving images about on the screen, as you will see in the example below. The second uses the 'canvas' object as a sort of whiteboard on which to draw graphs.

You can see a simulation of the system at www.essdyn.com/sim/ex1.htm. You will see two text-boxes which you can edit to change the parameters or the initial conditions.

11.11   Vibration with a Forcing Function

Up to this point we have assumed that the system is disturbed to start with, but then settles with no further disturbances. But it is likely that there will be some sinusoidal disturbance that is causing the vibration that we want to reduce. This could well be the result of a poorly balanced rotor in a spinning motor.

So we are particularly interested in the result of the system

$$x'' + a\,x' + b\,x = f(t)$$

when $f(t)$ is sinusoidal, such as $\cos(\omega t)$.

Now we first find the 'complementary function', the solution of

$$x'' + a\,x' + b\,x = 0$$

This is just the response of the system to a disturbance, as we have already found out.

But if we know that disturbances die away and we only want the long-term solution, we just have to find one function that will the 'particular solution' in response to that input $f(t)$.

If $f(t) = u\,\cos(\omega t)$ then it is the real part of $u\,e^{j\omega t}$.

We can assume that in response x is also proportional to $e^{j\omega t}$ and hence that

$$x' = j\omega\,x$$

and

$$x'' = -\omega^2\, x.$$

Our differential equation

$$x'' + a\, x' + b\, x = f(t)$$

becomes

$$-\omega^2 x + a\, j\omega\, x + b\, x = u\, e^{j\omega t}$$

so

$$x = u\, e^{j\omega t} / (-\omega^2 + a\, j\omega + b)$$

We can 'rationalise' this by multiplying top and bottom by $(b - \omega^2) - a\, j\omega$ to get

$$x = u\, e^{j\omega t} * (b - \omega^2 - a\, j\omega) / \{(b - \omega^2)^2 + a^2 \omega^2\}$$

so

$$x = u\{(b - \omega^2)\cos(\omega t) + a\, \sin(\omega t)\} / \{(b - \omega^2)^2 + a^2 \omega^2\}$$

Not altogether pretty, but a straightforward bit of algebra.

But if we now consider what happens if we excite the system at its undamped resonant frequency, we see that the real parts cancel out to leave

$$x = (u/a\omega)\, \sin(\omega t).$$

The amplitude is inversely proportional to the damping while the phase-shift is a ninety degree 'lead'.

To try this out with some numbers, let us add a force to our trolley example

Fig. 11.8 Trolley with damping and forcing function

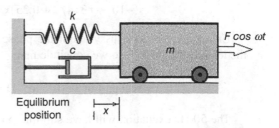

Before we put the numbers in, we see that in our equation

$$x'' + a\,x' + b\,x = f(t)$$

$$a = c/m$$
$$b = k/m$$
$$u = F/m$$

Exercise 11.1

If the coefficients are $m = 1000$ kg, $k = 250$ N/m, $c = 500$ Ns/m, $F = 10$ N,

1. What is the undamped resonant frequency?
2. What is the amplitude if the disturbance $\omega = 100$ pi rad/s?
3. What is the amplitude if the same magnitude force is applied at the resonant frequency?

Solution

Our equation

$$x'' + a\,x' + b\,x = f(t)$$

becomes

$$x'' + 0.5\,x' + 0.25\,x = 0.01\,\cos\omega t$$

The 'undamped natural frequency' is now 0.5 rad/s, the square root of 0.25.

When we excite the input so that both input and x are proportional to $e^{j\omega t}$, we have

$$(-\omega^2 + 0.5\,j\omega + 0.25)x = 0.01\,e^{j\omega t}$$

So if the excitation is at 50 Hz, giving $\omega = 100\,\pi$, then if we approximate π to $\sqrt{10}$ we have $\omega^2 = 10^5$ and so

$$(-10^5 + j\,314/2 + 0.25)x = 0.01\,e^{j\omega t}$$

Unless we are very fussy about the phase, we can ignore everything on the left except the 10^5, so we conclude that

$$x = \text{real part of } (-10^{-7}\,e^{j\omega t}) = -10^{-7}\,\cos\omega t$$

The 50 Hz excitation will have sub-micron effect.

But at the resonant frequency of 0.5 Hz it is a different story. Now the $-\omega^2$ will cancel out the 0.25, to leave

$$(-0.5*j*0.5)x = 0.01 \ e^{j0.5t}$$

so

$$x = \text{real part of } (0.04 \ j \ e^{j0.5t}) = 0.04 \sin 0.5t$$

The amplitude is 40 mm.

11.12   Base Excitation

A typical example that is used is that of a vehicle suspension or an antivibration mounting.

Fig. 11.9 Antivibration mounting

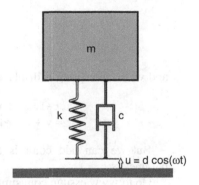

In this case, the forcing function depends on u and its derivative.

$$x'' + (c/m)(x' - u') + (k/m)(x - u) = 0$$

i.e.

$$x'' + (c/m)x' + (k/m)x = (c/m)u' + (k/m)u$$

and just as before we can substitute $e^{j\omega t}$ for $\cos(\omega t)$ and then perform some simple arithmetic with complex numbers to find the solution.

This example appears quite often in academic books, sometimes in an even more impractical form where the damper is shown to be attached to some fixed, non-vibrating point so that the u' term can be ignored. In fact that example is right here!

Example 11.3 Simulation of forced vibration

Fig. 11.10 Forced vibration

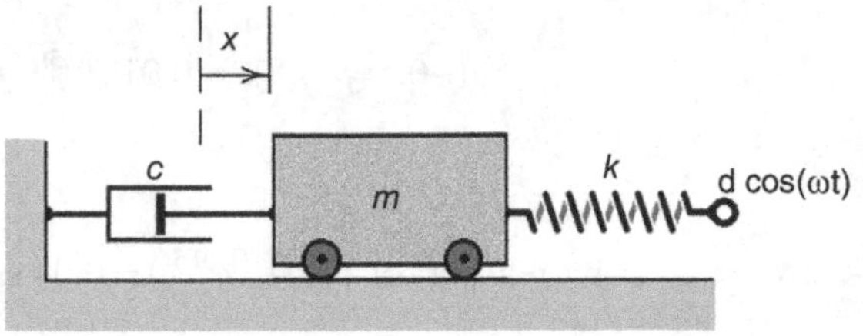

With a small time-step we can use simple Euler integration, where we say that the new displacement x is changed by adding the velocity v times the time-step dt.

```
x = x + v*dt;
```

Similarly the velocity is changed by adding to it the acceleration times dt.

For a system where

$$m = 1 \, kg$$
$$k = 25 \, N/m$$
$$c = 2 \, Ns/m$$

and where the spring is displaced by an input, we have

```
acceleration = 25*(input -x) - 2*v;
v = v + acceleration * dt;
```

But we can add controls to the simulation so that we can vary the parameters.

Go to www.essdyn.com/sim/resonance.htm to see the model in action.

You can set the damping to various values, including 0, and see the effect of the oscillatory input.

If you keep the stiffness at 25, set the damping to 0 and then increase omega to 5 you will see the vibration build up indefinitely – it steadily goes berserk.

If you increase the damping, the damping factor ζ will be (damping/10) for this stiffness, i.e. damping/$2\omega_n$. You will see that the output amplitude heads for a value that is equal to input times ω/damping-term, since at resonance the omega-squared term cancels out the constant term in the equation.

A bit of algebra will show you that this is input/2ζ.

1. Test the influence that changing the stiffness and damping values have for a given driving frequency, omega.

2. Test the influence that changing the driving frequency, omega, has for a given combination of stiffness and damping.
3. Now, consider the case where a machine starts from rest and thus omega is steadily increasing. Run the model and continually change omega in increments to see the effect of 'running through resonance'. Pay particular attention to the change in phase that occurs as you pass through resonance! (Here omega = 5, unless you have changed the stiffness.)
4. Set the damping factor to 0. Theory tells you that the response at the resonant frequency will have an infinite amplitude. What really happens?

Example 11.4 Vehicle suspension
The configuration below is the one shown by many textbooks. Although everything looks fine when the input is sinusoidal, there is trouble if we consider a step input. A shock-absorber suspension should be able to sustain a shock!

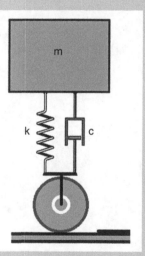

Fig. 11.11 A shocking suspension

If at speed the wheel hits a step, however small, it will cause an impulse to be delivered to the mass, because it takes infinite force to cause the damper to shrink at an infinite rate in zero time.

To make the damper shrink by an amount d will require an impulse dc Newton seconds to be applied. This will instantaneously cause the mass to gain a velocity of dc/m m/s. Something is likely to break!

Any practical shock-absorber must include another spring that acts in series with the damper.

Chapter 12
Modes

Abstract You have seen how spotting a pair of state variables can enable you to analyse a single-degree-of-freedom system that involves just a single frequency.

An example of such a system is a simple pendulum, swinging from side to side. But that pendulum has another degree of freedom where it swings towards us or away, or any combination of the two, swinging diagonally or in circles or ellipses. It has two frequencies, although in this case they are the same.

In general there will be numerous modes. Consider the system of a plate hanging on two springs that you met in the previous chapter.

As you well know, the mass has six degrees of freedom – three of position and three more of rotation. So there will be six ways in which it can vibrate or swing.

These are:

1. Up and down bouncing on the springs;
2. Rotary twisting on the springs about an axis 'into the page';
3. Sideways swinging;
4. Rotation about a vertical axis;
5. A combination of swinging towards us and away, where the plate tilts in the same direction as the swing;
6. Another combination where the tilt opposes the swing.

These can all happen at the same time!

Each of the first four can be analysed individually, using the techniques you have used for single degree of freedom systems.

But the last two require something more special.

12.1 Introduction

An example that is easier to visualise concerns two trolleys connected by springs.

Fig. 12.1 Trolleys and springs

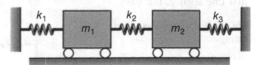

© Springer International Publishing AG 2018

J. Billingsley, *Essentials of Dynamics and Vibrations*,

DOI 10.1007/978-3-319-56517-0_12

121

Here it is easy to see that there are two **modes**. In one the trolleys move in unison, in the other they bounce together and apart.

You can see simulations in action:

At www.essdyn.com/sim/modesnodamp1.htm you can see one mode.

At www.essdyn.com/sim/modesnodamp2.htm you can see another.

At www.essdyn.com/sim/modesnodamp3.htm you can see how the two modes combine.

At www.essdyn.com/sim/modesnodamp4.htm you can see what happens if the middle spring is weakened, so that the modal frequencies are close together.

So how can we unscramble them?

First we must identify the 'state variables' – x_1, v_1, x_2, v_2 in this case.

Then we must write an equation for the rate-of-change of each of these.

Then we must work out the differential equation (fourth order in this case) that defines the motion and solve for a set of roots of the 'characteristic polynomial'.

The 'by the book' way to do this is to express the equations in matrix form and then look for 'eigenvalues' and 'eigenvectors' – you will see many details of this below.

If there is no damping, the 'fourth order' equation simplifies to a quadratic in the square of lambda. We can simplify the analysis by considering second-order equations, not first, and find eigenvalues that relate to the square of the oscillation frequency.

If there is damping, the roots will fall into complex conjugate pairs, representing a sine-wave multiplied by an exponential.

If all this mathematics worries you, do some urgent revision!

Only the simplest systems vibrate in just one 'mode'. Even the simple pendulum has two modes – it can swing left and right, or it can swing towards you and away. The frequencies of these modes are the same, but not their phases. If the two modes are 'in phase' the pendulum will swing diagonally, still in a plane. But if they are 90° apart, it will swing in a circle or ellipse.

We can make the frequencies different by arranging the string in a 'Y', suspended from two points with the strings joined part-way down. Then the bob will swing in a 'Lissajous' pattern. But the problems that you will encounter are more likely to involve masses and springs than a pendulum making a fancy pattern.

12.2 Springs and Masses

With one spring and a trolley, we worked out a simple second-order differential equation. But now we will add a second trolley. For symmetry we will add not just one spring, but two.

Fig. 12.2 Two trolleys, three springs

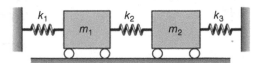

You will already have seen a simulation of the modes by following the links in the introduction.

But we need to be able to analyse what is going on.

As usual, our first action is to look for some equations of motion. But since there are no dampers, we can look for second order equations rather than first.

Now:

acceleration of mass $1 = ($compression of spring 1 plus extension of spring 2$)/m1$

acceleration of mass $2 = ($compression of spring 2 plus extension of spring 3$)/m2$

This gives

$$\ddot{x}_1 = [k_1(-x_1) + k_2(x_2 - x_1)]/m_1$$
$$\ddot{x}_2 = [k_2(x_1 - x_2) + k_3(-x_2)]/m_2$$

We can rearrange these to get

$$\ddot{x}_1 = [-(k_1 + k_2)x_1 + k_2 x_2]/m_1$$
$$\ddot{x}_2 = [k_2 x_1 - (k_2 + k_3)x_2]/m_2$$

which in matrix form becomes

$$\begin{bmatrix} \ddot{x}_1 \\ \ddot{x}_2 \end{bmatrix} = \begin{bmatrix} -\frac{k_1 + k_2}{m_1} & \frac{k_2}{m_1} \\ \frac{k_2}{m_2} & -\frac{k_2 + k_3}{m_2} \end{bmatrix} \begin{bmatrix} x_1 \\ x_2 \end{bmatrix}$$

For now let us say that the spring stiffnesses are all equal to k and that both masses are equal to m. We then have

$$\begin{bmatrix} -\frac{2k}{m} & \frac{k}{m} \\ \frac{k}{m} & -\frac{2k}{m} \end{bmatrix}$$

In the single-mode case, eigenvalues and eigenvectors gave the answer. So let us find the eigenvalues of

$$\begin{bmatrix} -\frac{2k}{m} & \frac{k}{m} \\ \frac{k}{m} & -\frac{2k}{m} \end{bmatrix}$$

Now

$$det \begin{vmatrix} -\frac{2k}{m} - \lambda & \frac{k}{m} \\ \frac{k}{m} & -\frac{2k}{m} - \lambda \end{vmatrix} = 0$$

so

$$\lambda^2 + 4\frac{k}{m} + 3\frac{k^2}{m^2} = 0$$

This has roots $\lambda = -k/m$ and $\lambda = -3k/m$.
When we substitute these values in turn into

$$\begin{bmatrix} -\frac{2k}{m} - \lambda & \frac{k}{m} \\ \frac{k}{m} & -\frac{2k}{m} - \lambda \end{bmatrix} \begin{bmatrix} \xi_1 \\ \xi_2 \end{bmatrix} = \begin{bmatrix} 0 \\ 0 \end{bmatrix}$$

we find that the eigenvectors are $(1, 1)'$ and $(1, -1)'$.
    Now λ will represent $-\omega^2$ in this undamped second-order case, so

$$\omega = \sqrt{(k/m)} \text{ or}$$
$$\omega = \sqrt{(3k/m)}$$

But what does it mean?
    Using the convention of representing sines and cosines as complex exponential, we can express the solution as

$$\begin{bmatrix} x_1 \\ x_2 \end{bmatrix} = A \begin{bmatrix} 1 \\ 1 \end{bmatrix} e^{j\sqrt{\frac{k}{m}}t} + B \begin{bmatrix} 1 \\ -1 \end{bmatrix} e^{j\sqrt{\frac{3k}{m}}t}$$

    A and B are complex numbers that define the amplitude and phase and then we take the real part. We find A and B from the initial conditions.
    As we saw in the simulation, the motion can be made up of either of the modes or a mixture of the two.

12.2.1   Note

In the analysis above, we have divided the equations through by the masses, to obtain a clean second derivative of each x with a unit coefficient. But in many textbooks you will see the equations in the form

$$[M]\{\ddot{x}\} + [K]\{x\} = \{0\}$$

This means that you will be trying to solve

$$([K] + \lambda^2[M])\{X\} = \{0\}$$

so to consider the solution in eigenvector terms you have to worry about calculating $\mathbf{M^{-1}K}$.

Admittedly the inverse of $\mathbf{M}$ is just a diagonal matrix with elements $1/m_i$, but surely it is easier to incorporate the masses when forming the equations in the first place.

12.2.2 Summary

In general if there is no damping the method is as follows.

1. Derive second-order differential equations for the positions of each of the masses. These will involve the positions of some or all of the other masses.
2. Express these equations in matrix form.
3. Find the eigenvalues λ_i of the matrix.
4. Find the eigenvectors $\boldsymbol{\xi}_i$ corresponding to the eigenvalues.
5. Calculate the (angular) frequencies $\omega_i = \sqrt{(-\lambda_i)}$
6. The vector of displacements of the masses will be given by

$$\mathbf{x} = \Sigma A_i \boldsymbol{\xi}_i e^{j\omega_i t}$$

12.3 Other Types of Oscillation

Example 12.1

You should now be familiar with a pendulum of length L.

Now make it a double pendulum by adding a further string of length L and an equal mass to hang below it.

What is the ratio of the two modal frequencies?

Fig. 12.3 Double pendulum

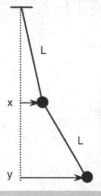

Solution

For the simple pendulum we had

$$x'' = -g/L\,x$$

where the deflection was x, so that the angle of the string was x/L.

Now we have to consider the deflection of the lower mass, y.

It is hanging from the top mass, so the angle of the string is $(y-x)/L$

$$y'' = g/L(x-y)$$

But now the top string has tension 2 mg, while the lower string is also pulling the top mass with 1 mg.

So

$$x'' = -2g/L\,x + g/L(y-x)$$

i.e.

$$x'' = -3g/L\,x + g/L\,y$$

We have the information we need to make a matrix equation

$$\begin{bmatrix} \ddot{x} \\ \ddot{y} \end{bmatrix} = \begin{bmatrix} -3\frac{g}{L} & \frac{g}{L} \\ \frac{g}{L} & -\frac{g}{L} \end{bmatrix} \begin{bmatrix} x \\ y \end{bmatrix}$$

and look for eigenvalues.

The characteristic equation, given by det $|A - \lambda I| = 0$, is

$$(\lambda + 3g/L)(\lambda + g/L) - g^2/L = 0$$

i.e.

$$\lambda^2 + 4g/L\,\lambda + 2g^2/L^2 = 0$$

so

$$\lambda = (-4 \pm \sqrt{(16-8)})g/2L$$

$$\lambda = (-2 \pm \sqrt{2})g/L$$

The ratio of the frequencies will be the square root of $(2 + \sqrt{2})/(2 - \sqrt{2})$

Now in the same way that we rationalise complex numbers, we can multiply top and bottom by $(2 + \sqrt{2})$ to get

$$\text{Ratio} = (2 + \sqrt{2})^2 / (4 - 2).$$

So its square root is

$$(2 + \sqrt{2})/\sqrt{2} = 1 + \sqrt{2}.$$

The ratio of the frequencies is 2.414 to 1.

Check if this matches the simulation at www.essdyn.com/sim/double.htm.

Example 12.2

In Example 11.1 you were presented with a hanging plate, that you only considered to move vertically.

The springs are 2 m apart and the plate is 4 m wide by two high. So the principal moments of inertia are $J_x = m/3$, $J_y = 5m/3$ and $J_z = 4m/3$, where z is 'up' and y is 'into the paper'.

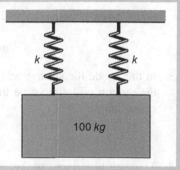

Fig. 12.4 Hanging plate

As you well know, the mass has six degrees of freedom – three of position and three more of rotation. So there will be six ways in which it can vibrate or swing. These are:

1. Up and down bouncing on the springs
2. Rotary twisting on the springs about an axis 'into the page'
3. Sideways swinging
4. Rotation about a vertical axis
5. A combination of swinging towards us and away, where the mass tilts in the same direction as the swing
6. Another combination where the tilt opposes the swing.

These can all happen at the same time!

Each of the first four can be analysed individually, using the techniques you have used for single degree of freedom systems in the last chapter.

But now you have the techniques needed to analyse the last two.

Solution

Let us consider these modes

1. We have already covered the first mode in the previous chapter.
2. For rotation about the plate's centre, our state variables will be the angle θ and its rate.

    We know the moment of inertia J_y about the axis perpendicular to its centre

$$J_x = 5\,m/3.$$

   We know the extensions that the rotation θ will make to the springs, 1 metre times θ to one of them and $-1\,\theta$ to the other, and so can work out that the restoring couple is $-2k\theta$. So we arrive at the differential equation

$$J_y\theta'' = -2k\theta$$

So the angular frequency will be given by

$$\omega = \sqrt{(2k/J_y)}$$

3. In this mode the springs will not stretch and the plate will not rotate, so the motion will simply be that of a pendulum with length equal to the spring length L.

$$\omega = \sqrt{(g/L)}$$

4. This is a little more tricky. Again the springs will not stretch. Since the springs are attached 1 metre each side of the centre, a rotation angle θ will cause the springs to be slanted at an angle $1 \times \theta/L$ to the vertical. This will cause a restoring couple $mg\theta/L$ – since there are two springs, but each carries only half the weight. So now

$$J_z\theta'' = -mg\theta/L$$

and since we know that $J_z = 4m/3$

$$\theta'' = 3g/4L$$

$$\omega = \sqrt{(3g/4L)}$$

5 and 6. These call for something different. We have a pair of modes acting
together and will have to define two angles, that of the springs, θ,
and that of the plate, φ.

The tension $mg/2$ in each spring will have a horizontal component $mg\theta/2$.
Together they will accelerate the centre of mass y of the plate

$$y'' = -g\theta$$

The springs will make an angle $(\theta - \Phi)$ with the plate, so by taking
moments about its centre of mass we will get

$$J_x\varphi'' = mg(\theta - \varphi)$$

Since $J_x = m/3$ we can write this as

$$\varphi'' = 3g(\theta - \varphi)$$

Now we must express y in terms of θ and φ

$$y = L\theta + 1\varphi$$

so the equation for y'' becomes

$$L\theta'' + \varphi'' = -g\theta$$

and when we substitute for φ'' we get

$$L\theta'' + 3g(\theta - \varphi) = -g\theta$$

i.e.

$$L\theta'' = -4g\theta + g\varphi$$

so

$$\theta'' = -4g\theta/L + g\,\varphi\,L$$

So at last we have two equations from which to make a matrix equation

$$\begin{bmatrix} \ddot{\theta} \\ \ddot{\phi} \end{bmatrix} = \begin{bmatrix} -4\frac{g}{L} & \frac{g}{L} \\ 3g & -g \end{bmatrix} \begin{bmatrix} \theta \\ \phi \end{bmatrix}$$

When we look for the eigenvalues we get a characteristic equation

$$(\lambda + 4g/L)(\lambda + g) - 3g^2/L = 0$$

i.e.

$$\lambda^2 + (4/L + 1)g\lambda + g^2/L = 0$$

Now if we know L, we can work out the modal frequencies. Suppose that we take $L = 1$, then

$$\lambda^2 + 5g\lambda + g^2 = 0$$
$$\lambda = (-5 \pm \sqrt{(25-4)})g/2$$
$$\lambda = (-5 \pm \sqrt{21})g/2$$

But now $\omega = \sqrt{(-\lambda)}$, so we have a little more calculation to do if we want a numerical answer.

12.4   State Equations and Damping

When there is damping, we can no longer rely on the 'simple harmonic motion' equations that do not include velocity. For our two-mass system we will have four variables, not two, being the two displacements and the two velocities.

We can define v_1 as dx_1/dt and $v_2 = dx_2/dt$, giving us two of our four state equations straight away:

$$\dot{x}_1 = v_1$$
$$\dot{x}_2 = v_2$$

so that the matrix equation is now a first-order one

$$\begin{bmatrix} \dot{x}_1 \\ \dot{v}_1 \\ \dot{x}_2 \\ \dot{v}_2 \end{bmatrix} = \begin{bmatrix} 0 & 1 & 0 & 0 \\ a & b & c & d \\ 0 & 0 & 0 & 1 \\ e & f & g & h \end{bmatrix} \begin{bmatrix} x_1 \\ v_1 \\ x_2 \\ v_2 \end{bmatrix}$$

where a to h are constants that you derive from the springs and dampers.

You can see that the characteristic equation is now of fourth order. It can be factorised into two quadratics, each of these representing its own undamped natural frequency and damping factor.

So what is important about these quadratics?

Each will represent a second order system of the form

$$\ddot{x} + 2a\dot{x} + bx = 0$$

(these are not the same a and b as the ones above) which is often be written in a more complicated way as

$$\ddot{x} + 2\zeta\omega_n\dot{x} + \omega_n^2 x = 0$$

where the symbols 'zeta' and omega$_n$ have some special significance as 'damping factor' and 'undamped natural frequency'.

If the damping is small, the solution will represent a sine wave 'modulated' by a decaying exponential. It is this exponential that will interest you, since it determines how fast any disturbance will die away.

When you solve the quadratic for the roots, you will find that the real part of the pair of complex roots is just

$$-a$$

so that middle term gives you the coefficient of the exponential decay. You will want it to be large and negative for a fast decay.

As you increase the damping the resonant frequency will decrease. At critical damping you will only have a decay with no sine wave, but with a multiplying time t thrown in to one of the terms.

Beyond this, one of the real roots will diminish while the other increases.

So what about TWO quadratics? How do you choose the best damping when both will change? Well it is obviously the slower of the two modes that will bother you. You do not mind if the other one gets slower, as long as it remains the faster.

Exercise 12.1

The two-trolley three-spring system has been modified by doubling the second mass to 2 kg. The springs are now of stiffness 10 N/m.

Fig. 12.5 Trolleys, springs and damper

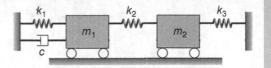

The system to be analysed is simulated at www.essdyn.com/sim/modesq.htm

1. Calculate the new characteristic equation if there is no damping, using a two-by-two matrix.
2. Instead express the equations in first-order form as a four-by-four matrix, where the damper value is c Ns/m.

3. What is this new characteristic polynomial? (Hint: if you set $c = 0$, you should have a similar expression to 1, but with the lambdas squared.)
4. Using trial and error on the model, for what value of c will the oscillations die away fastest?

Solution
Ignoring the damper, we will have two second-order equations that can be expressed in matrix form as

$$\begin{bmatrix} \ddot{x}_1 \\ \ddot{x}_2 \end{bmatrix} = \begin{bmatrix} -\frac{k_1+k_2}{m_1} & \frac{k_2}{m_1} \\ \frac{k_2}{m_2} & -\frac{k_2+k_3}{m_2} \end{bmatrix} \begin{bmatrix} x_1 \\ x_2 \end{bmatrix}$$

as we saw in Section 2.

When we substitute the new values of $k_1 = k_2 = k_3 = 10$ N/m and $m_1 = 1$ kg, $m_2 = 2$ kg we get

$$\begin{bmatrix} \ddot{x}_1 \\ \ddot{x}_2 \end{bmatrix} = \begin{bmatrix} -20 & 10 \\ 5 & -10 \end{bmatrix} \begin{bmatrix} x_1 \\ x_2 \end{bmatrix}$$

with a characteristic equation

$$\lambda^2 + 30\lambda + 150 = 0$$

and roots

$$\lambda = -15 \pm \sqrt{(225 - 150)}$$

i.e.

$$\lambda = -15 \pm 5\sqrt{3}$$

But with the damping,

$$x_1'' = -20\,x_1 + 10\,x_2 - c\,x_1'$$

so when we want to include the damping, we have to use a four-variable first-order equation

$$\begin{bmatrix} \dot{x}_1 \\ \dot{v}_1 \\ \dot{x}_2 \\ \dot{v}_2 \end{bmatrix} = \begin{bmatrix} 0 & 1 & 0 & 0 \\ -20 & -c & 10 & 0 \\ 0 & 0 & 0 & 1 \\ 5 & 0 & -10 & 0 \end{bmatrix} \begin{bmatrix} x_1 \\ v_1 \\ x_2 \\ v_2 \end{bmatrix}$$

Now the characteristic equation will require somewhat more calculation

$$det \begin{vmatrix} -\lambda & 1 & 0 & 0 \\ -20 & -c-\lambda & 10 & 0 \\ 0 & 0 & -\lambda & 1 \\ 5 & 0 & -10 & -\lambda \end{vmatrix} = 0$$

A little effort results in

$$\lambda(\lambda + c)(\lambda^2 + 10) + 20(\lambda^2 + 10) - 50 = 0$$

i.e.

$$\lambda^4 + c\lambda^3 + (20 + 10)\lambda^2 + 10c\lambda + 200 - 50 = 0$$

giving

$$\lambda^4 + c\lambda^3 + 30\lambda^2 + 10c\lambda + 150 = 0.$$

Sure enough, if $c = 0$ then the equation is the same as before, except that we have λ^2 in place of λ.

But solving this will now call for some computer assistance!

If we try $c = 5$ we will be trying to factorise

$$\lambda^4 + 5\lambda^3 + 30\lambda^2 + 50\lambda + 150 = 0$$

An online factoriser gives

$$(x^2 + .874x + 8.28)(x^2 + 4.126 + 18.112)$$

The roots are approximately $-2 \pm 3.7j$ and $-0.44 \pm 2.8j$.

So we see that the slower mode will die away with a time-constant of just over 2 seconds, while the faster one will die away with a time constant of half a second. Play with the simulation parameters to see if you can do better.

Exercise 12.2
To finish the section, here is an example that will really try your strength. In the previous chapter I grumbled about a suspension that would not absorb a shock, stating that a second spring was required. Well here it is.

A simplified suspension for a railway carriage wheel is shown below.

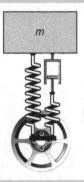

Fig. 12.6 Damped suspension

Without the second spring, the suspension would transfer shocks to the vehicle.

The mass of the wheel itself can be ignored.

The inertial load supported by that wheel can be assumed to represent a point mass of 300 kg.

The spring that supports the weight has stiffness 10,000 N/m.

The train travels at speed, experiencing variations of track surface height $u(t)$.

(a) Without the damper and its spring, derive a second order differential equation to describe the vertical deflection x of the carriage.
(b) What will be the frequency of the undamped oscillations? (cycles per second)The shock-absorber is added as shown, in the form of a spring of stiffness 80,000 N/m in series with a damper of coefficient c Ns/m.
(c) Two of the state variables are the vertical deflection x and its rate-of-change $\dot{x}$. Show that a third variable must be added to describe the motion. What is it?
(d) Derive three first-order differential equations to describe the motion.
(e) Represent these equations in matrix form, involving c as a parameter and $u(t)$ as an input.
(f) Derive the characteristic equation to find the eigenvalues. (Hint: Do not be too quick to convert fractions to decimals.)
(g) What is significant about the response if $c = 8000/3$ Ns/m? Explain.

Solution
(a) Without the damper and its spring, derive a second order differential
equation to describe the vertical deflection x of the carriage.
   The compression of the spring is $(u(t) - x)$, so

$$300\ddot{x} + 10000x = 10000f(t)$$

(b) What will be the frequency of the undamped oscillations? (cycles per
second)

$$\omega = \frac{10}{\sqrt{3}}, \ \text{so} f = \frac{4}{\pi\sqrt{3}}$$

which is 0.92 cycles per second.
The shock-absorber is added as shown, in the form of a spring of stiffness 80,000 N/m in series with a damper of coefficient c Ns/m.
Now it gets more interesting.
(c) Two of the state variables are the vertical deflection x and its rate-of-
change $\dot{x}$. Show that a third variable must be added to describe the
motion. What is it?
   To work out the compression of the second spring, we have to know
the length of the damper.
   So the three state variables are x, v and y, where v is $\dot{x}$ and y the
length of the damper.
(d) Derive three first-order differential equations to describe the motion.
   The first spring is compressed by an amount $(u - x)$
   The second spring is compressed by an amount $(u + y - x)$
   So the force on the damper is $80000 \times (u + y - x)$
while the total force on the mass is $10000 \times (u - x) + 80000 \times (u + y - x)$
The three equations become:
1. $dx/dt = v$

2. $300\, dv/dt = 10000 * (u - x) + 80000 * (u + y - x)$
    $= 90000 * u - 90000 * x + 80000 * y$

3. $c * dy/dt = -80000 * (u + y - x)$

(e) Represent these equations in matrix form, involving c as a parameter
and $u(t)$ as an input.
   We can tidy up these equations to get

$$\begin{bmatrix} \dot{x} \\ \dot{v} \\ \dot{y} \end{bmatrix} = \begin{bmatrix} 0 & 1 & 0 \\ -300 & 0 & \frac{800}{3} \\ \frac{80000}{c} & 0 & \frac{-80000}{c} \end{bmatrix} \begin{bmatrix} x \\ v \\ y \end{bmatrix} + \begin{bmatrix} 0 \\ 300 \\ \frac{-80000}{c} \end{bmatrix} u(t)$$

(f) Derive the characteristic equation to find the eigenvalues. (Hint: Do not be too quick to convert fractions to decimals.)

Now we stick in some lambdas and take the determinant:

$$\det \begin{vmatrix} -\lambda & 1 & 0 \\ -300 & -\lambda & \frac{800}{3} \\ \frac{80000}{c} & 0 & \frac{-80000}{c} - \lambda \end{vmatrix} = 0$$

Expanding by the top row:

$$-\lambda * (-\lambda) * (-\lambda - 80000/c) - 1 * \{(-300) * (-\lambda - 80000/c) \\ - (80000/c) * (800/3)\} = 0$$

This will be neater if we multiply through by -1:

$$\lambda^3 + 80000 * \lambda^2/c + 300 * \lambda + 300 * 80000/c - 800 * 80000/(3 * c) = 0$$

which boils down to

$$\lambda^3 + \frac{80000}{c}\lambda^2 + 300\lambda + \frac{100 * 80000}{3c} = 0$$

(g) What is significant about the response if $c = 8000/3$ Ns/m? Explain. Nearly there!

When we substitute that value for c, we get

$$\lambda^3 + 30\lambda^2 + 300\lambda + 1000 = 0$$

which should be easy to spot as

$$(\lambda + 10)^3 = 0$$

There are three equal real roots of 0.1 second, a sort of 'ultracritical' damping!

12.5 Conclusion

My earnest hope is that now you have a full understanding of the principles, so that you can tackle new problems, not ones that are just formulated as something that you can solve by applying a formula that you have learned.

Here is an exercise that is just a little different. I will not give a solution. If you cannot find the solution yourself you should really study the chapter again.

Exercise 12.3
Figure 12.7 shows two frictionless trolleys linked and connected to a base
by two springs.

Fig. 12.7 A slightly different
problem

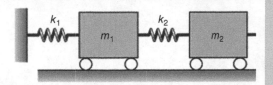

The springs both have stiffness 10 kN/m.
The two modes have frequency $1/2\pi$ and $1/\pi$ Hz.

(a) What is the mass of each of the trolleys?
(b) What are the eigenvectors of the modes?

Approximate π^2 to 10.

Chapter 13
Rocket Science

Abstract This chapter pokes fun at the misconceptions of the media industry. From a tender age, you have been exposed to the cinema film director's concepts of dynamics – and so many of them are wrong! When a 'starfighter' banks, it is shown zooming away in a curve. But Newton's laws tell you that you have to fire a jet perpendicular to the path if you want to change its line. The star-ship 'Enterprise' is shown with the unfortunate property that if its engines fail, its orbit will decay rapidly. But it is only the satellites that are in an orbit so low that air drag takes effect that need have any worry. It is not only the fiction writers who are to blame. Many documentaries have blunders that are far less forgivable. If the orbit of a satellite starts to decay, it can be boosted into a higher orbit by propelling it to a higher velocity. A UK documentary showed the jets propelling the satellite straight up. Is this really the correct answer? Some literary critics hail Jules Verne as a scientific genius. But his books are full of mathematical and engineering howlers. Jules Verne has the excuse of having written his books long ago, but the same cannot be said for the perpetrators of the film 'Gravity' that breaks almost every kinetic law in the book. When it comes to the mythology that surrounds 'black holes' all common sense is lost. Every newsreader seems to report that black holes are sucking everything in like a giant vacuum cleaner, while other writers see them as the ends of wormholes.

13.1 Introduction

In the middle of the last century, the film 'Destination Moon' strove to be accurate in every detail. I was young when I watched it, but I cannot remember any errors. A cartoon sequence near the start even explained the physics of rocket propulsion. Its shortcoming was a reliance on a nuclear reactor to boil water to a high enough temperature to be a credible reaction mass.

At around the same time, a film 'Rocketship XM' was full of blunders. The cabin was gymballed in order to 'take advantage of any gravity' – the whole concept of free fall was lost on the producer. As a teenager I became aware of the liberties that were being taken with dynamics.

J. Billingsley, *Essentials of Dynamics and Vibrations*,
DOI 10.1007/978-3-319-56517-0_13

13.2   Black Holes and Gravity Gradients

In 1966 I attended an ESRO Summer School on the stabilisation of spacecraft. It was not as dramatic as it sounds. Not many years after Sputnik, reaction mass was very expensive to haul into orbit. Instead of using jets to control the attitude of the satellites, for the Europeans 'gravity gradient' was the method of choice.

Long poles extended up and down from the body of the satellite. The inverse square law meant that the upward pole would weigh very slightly less than its downward counterpart. The forces involved were minuscule, but in free fall in space they were enough to set the satellite swinging, albeit very slowly, about an attitude that was vertical. The oscillation was damped by connecting the poles via hinges which included magnetic damping.

But when the Americans gave a presentation about their weather satellite, cranking the solar panels around almost a complete revolution every day, it seemed an obscene waste of energy!

So what has this got to do with Black Holes?

Suppose that the Earth collapsed to form a black hole. Its 'event horizon' would be about the size of a pebble. Now suppose that you fly your spaceship a thousand miles from that black hole, where the gravitational attraction will be 16 times that of the Earth's surface gravity 'g'. Will you come to any harm? Not a bit. Your spaceship would gain plenty of momentum on the inward trip to take it swooping safely out again, while you would still float in free fall within it, feeling none of that gravity.

If you tried to fly that close to the centre of the earth you would instead crash into the ground! So a black hole of that size would be much safer to navigate around than the equivalent planet.

Now suppose that you are adventurous and try to fly somewhat closer to the black hole. Would you be safe at a hundred miles? The gravitational attraction will then be 1600 g, but you could still zoom in and out, just the same.

But what about gravity gradient? It is minuscule at the height of a satellite's orbit, but how would you fare at a hundred miles from the black hole?

At radius r, the gravity will be gR^2/r^2, where R is the radius of the Earth. So its gradient will be the derivative, $-2gR^2/r^3$. The tug between your head and your legs, supposing that you curl up to keep them just one metre apart, will be g times 3200 times one metre divided by 100 miles, or about 3200/160,000 g, about a fiftieth of a gravity. You will hardly feel it.

But at one mile it is a different story. The gradient will be a million times as great and your head will be pulled from your body with a force of its mass times 20,000 g. If the spaceship has not also been pulled apart, your remains will be smeared at its two ends.

So much for the idea of travelling through a 'wormhole' that starts at a black hole!

It is rather hard to 'fall into' a black hole. Unless you hit it square on, your linear and angular momentum will carry you back out again. Your biggest risk, apart from gravity gradient is to hit another victim coming in the opposite direction.

Black holes pose some interesting conundrums. What happens to the entropy previously amassed by things that fall into them? Does its loss break that law of thermodynamics, that entropy always increases? But TV science journalists are apt to confuse thermodynamic entropy with the entropy of information theory.

Science journalists love an excuse to proclaim that a black hole is at the centre of a galaxy. Now the orbital velocities of satellites circling a concentrated mass, such as the Sun, will lie on an 'orbit curve'. Since g/r^2 must equal v^2/r, the square of the velocity will be inversely proportional to the radius. Now the velocity rises to infinity as zero radius is approached.

The galaxy in question had the property that the orbital velocities of the stars in its arms were more or less the same, for any radius. This was surely proof that there was NOT a significant black hole at its centre! (I think that the constant velocity property requires the stellar density to be greater near the centre, so that there is an equal mass of stars in a ring at any radius.)

13.3 Rockets Galore

Another documentary that caught my attention explained that a satellite could be boosted into a higher orbit by firing a jet vertically. In fact this would be disastrous! The new orbit would intersect the original orbit at an angle, and although the altitude would initially increase, when the satellite came around again it would descend much lower. Since the thrust is perpendicular to its velocity, its actual orbital energy would not have been increased and the orbit would just be made 'more elliptical'. The correct procedure is of course to propel the satellite to a higher velocity, causing it to have risen when it reaches the 'other side' of the orbit and another propulsive 'burn' can be given to enable it to remain at that new height.

Getting away from grumbles for a moment, space propulsion is full of contradictions. Why is it of advantage to send a craft to Mars via the 'Grand Tour' that first takes it 'downhill' to pass by Venus?

In ejecting 'reaction mass' at great velocity, a rocket motor can exert a force. But that does not necessarily mean that it imparts energy. A launch vehicle could hover a few feet above the ground until the fuel ran out, and still not get anywhere. It is really all about momentum and 'impulse'. Now remember that since energy is force times distance, the energy imparted to the spacecraft will be proportional to the velocity at which it is travelling at the time. So by descending to the orbit of Venus and speeding up even further as it falls into the Venus gravity-well, the burn can be made at maximum speed and the energy contribution of the thrust can be multiplied many times over. And of course the 'downhill' energy is still retained to help the craft bounce back out.

I researched control theory in Cambridge, where at the same time Stephen Hawking and I were Research Scholars of Trinity Hall. During that time a distinguished Russian professor came to visit the Control Group. He gave a presentation on his derivation of a fuel-optimal strategy for an unmanned lunar landing. Now at that time, two of the Soviet moon-shots had crashed.

The fuel has to counter the terminal velocity of falling from space onto the Moon. But it also has to make up for the momentum gained by the vehicle during the time that it is descending. So the fuel-optimal strategy is to descend as fast as possible. The rocket motor is switched on at the last possible moment, applying full thrust until the velocity just reaches zero as the vehicle touches the surface.

I delicately pointed out to my professor that since the craft might be descending at a mile a second when the motor is turned on, if the burn starts one second late then the vehicle would end up one mile underground – if the Moon's surface did not get in the way first.

My research was on suboptimal control and I suggested that a suboptimal strategy might be a lot safer, since a consequence of minimising the fuel was to maximise the risk of failure. Left-over fuel was of no value on that one-way trip. The suboptimal strategy would mean timing the start of the burn as though the motor only had 90% of its thrust. Under full thrust the velocity would soon drop below 'expectation', at which the thrust could be reduced as the performance data became more precise. But there would be thrust to spare to compensate for errors, while the increase in the fuel used would be minimal.

My professor had a delicate word with the visiting professor and a few months later the next Soviet vehicle landed safely. So be very careful what you optimise!

13.4   The Gravity of Errors

In general, films can have the excuse that by inventing gravity generators in their spaceships, their actors can keep their feet on the ground in an earthbound film studio. (However all too often the storyline has somebody falling to their doom 'down' the length of the spaceship.) I have to admit that other films take pains to portray 'centrifugal' artificial gravity of the practical sort. I recall a set that was curved vertically to represent a 'spinning' ring habitat. The set could be tipped to keep the actors perpendicular to the 'floor'.

The film 'Gravity' has no such excuse, containing many scenes in which the actors float about. But the errors start with the whole theme of the film and continue throughout.

As I recall it, the leading characters are in an orbiting space station. At a different place in the orbit another satellite blows up and the concentrated wreckage hurtles towards them. Why should it?

A common fault in films of the 'Star Wars' genre is to show one craft pursuing another. When it fires and succeeds in making the leading vessel explode, it flies through a cloud of wreckage, just as it might in Earth's atmosphere. But of course there is no atmosphere in space. If both craft have the same velocity, the leading craft will explode in an expanding sphere of fragments with a centre that had no velocity relative to the pursuer.

In the same way, the exploding satellite in 'Gravity' might certainly project some fragments towards our heroes, but would never approach them in a concentrated mass. What is more, any mass with a substantial velocity relative to our heroes would not remain in that same orbit!

But the ultimate offence against physics comes when the two heroes are drifting from their punctured refuge towards another capsule in the same orbit. Clinging together they succeed in halting their travel by tangling in some parachute shrouds. The hero hangs onto the heroine's ankle, but nobly releases his grasp for fear that he might drag the heroine free. Heroically he accelerates away towards oblivion.

But what on earth (or in space!) makes him accelerate? With the capsule and both bodies in free fall, the tension of a single hair could accelerate him to drift to safety in the capsule.

If the capsule and ropes had been shown spinning, that part of the plot could have been made plausible, but adherence to the laws of physics was clearly something low in the priorities of the producers.

13.5 Jules Verne and Science Fiction

A characteristic of many of Verne's novels is a meticulous attention to numbers. In 'From Earth to Moon', although he realised that a large acceleration would be required to propel his astronauts to the moon by means of a gunshot, he seems to have overlooked its magnitude. He proposed a gun barrel 270 m in length to accelerate the passengers in his capsule to the 11 km/s escape velocity. That would involve an acceleration of some 20,000 g!

To cushion the acceleration he proposed springs and a collapsible floor, that would not in fact make a whisker of difference.

In 'Journey to the Centre of the Earth' he takes his protagonists to a depth of 142 km. Even if we ignore the hazards of molten magma, the air pressure would have been an insuperable obstacle. Pressure reduces by a factor of e for every 10 km that you ascend and increases accordingly when you descend. So at a depth of 142 km the pressure would be e to the power 14.2 atmospheres, some 1.5 million atmospheres, had the air not liquefied first.

Not all science fiction has been about exploration, space cowboys or monsters. In its heyday in the middle of the twentieth century there were speculative tales that ranged from personality transfer into a computer to novel economic systems! How would the economy survive if anything could be duplicated for nothing? Or with the ultimate in automated manufacturing, perhaps you could have anything you ask for, again for nothing. Robots were given personalities, good and evil, and in one series became the focus of detective stories.

Such stories have no misrepresentation of dynamics to offend the engineer, but could give the economists and sociologists food for thought.

As morality seems to be increasingly eroded, censorship is almost a thing of the past. But maybe we should ask for censorship to be brought back on material that is misleading in the fundamental facts of physics, or at the very least we should insist that some films should carry a credibility warning.

While increasing your understanding of dynamics, I hope that this book has not destroyed your enjoyment of space films!

The text of Jules Verne's novels can be downloaded, 'From Earth to Moon and a Journey Around it' from http://www.online-literature.com/verne/moon-voyage/ and 'Journey to the Centre of the Earth' from http://www.online-literature.com/verne/journey_center_earth/.

Appendix 1: Mathematicians and Operators

Abstract There is some rather formidable mathematics in the study of dynamics. In the chapters, some attempt has been made to lighten the load, asking the reader to take some assertions on trust. In these appendices the mathematics is allowed to come to the fore. This first appendix presents the use of partial differentiation to form operators that enable us to derive more far-reaching formulae such as Euler's equations.

A.1.1   Introduction

Mathematicians love operators. The one that you will be most familiar with is the **differential operator**, the simple d/dt. You look at the change in something over a small interval of time, divide by the length of that interval, and you have an estimate of the rate-of-change.

But the functions you deal with can involve other variables than time. For example the potential energy of an object depends on its height above the ground. In that case you might be interested in d/dz, where z is the height.

But take your object into space and gravity depends on all three coordinates, x, y and z. Now you are interested in the rate-of-change of potential if you change, say, x alone. The 'd' in the operator becomes the 'curly-d' of **partial differentiation**

$$\partial/\partial x$$

But there's more. All three partial derivatives can be bunched together into a vector, leading to yet another operator **grad**

$$\nabla = (\partial/\partial x, \partial/\partial y, \partial/\partial z)'$$

The triangle symbol is called 'del' or sometimes 'nabla', and when applied to a scalar gives a vector result, as when finding the gradient of a potential field.

Grad has two partners, **div** and **curl**, but these will be of more interest to electrical engineers and fluid dynamicists. The operator is the same, but is now used

© Springer International Publishing AG 2018
J. Billingsley, *Essentials of Dynamics and Vibrations*,
DOI 10.1007/978-3-319-56517-0

as the scalar (div) or vector (curl) product with a vector field, such as the velocity $\mathbf{v}(x, y, z)$ at each point of a moving fluid. Div measures the 'divergence' of the flow, the rate at which the fluid is expanding. If it is incompressible, for example water, we can write

$$\nabla \cdot \mathbf{v} = 0$$

On the other hand, curl gives a vector result. When you have stirred your coffee, $\nabla \times \mathbf{v}$ would tell you its angular velocity at that point.

But there are even more operators that you do have to consider. As soon as you change your frame of reference from something fixed to something rotating, for example looking at the forces you feel inside an aircraft that is rolling, the d/dt operator splits into one part that is the partial derivative with respect to time and another that is the cross product with the vector angular velocity

$$\frac{\mathrm{d}}{\mathrm{d}t} = \frac{\partial}{\partial t} + \omega \times$$

You will have to get used to a great many **vector** quantities. Moments, angular velocities and angular momentum are just a start. You will also have to meet **tensors**. Now a tensor is really just a matrix – though the mathematicians like to fatten it up to more than two dimensions when they can.

If I have a solid object, its inertia will be in the form of a tensor. If I give it an angular velocity omega, the angular momentum will be the product of the tensor with the vector omega. It will be yet another vector, and if the object is not nicely balanced it can be in a different direction from the angular velocity. As omega rotates, the momentum vector rotates and the bearings of whatever shaft is involved have to supply a couple to keep it changing.

And now we are entering the world of vibration.

One vibrating mass is easy, as long as you are happy with second order differential equations. But link a number of masses together with springs, and the problem escalates into a matrix of differential equations with eigenvectors defining the modes and eigenvalues defining the frequencies.

I warned you that it would not be easy! But the aim of these notes is to make things as easy to understand as possible.

A.1.2   Simple Time Derivative Versus Discrete Time Computation

You hold a ball above the ground at height z and let it go.

Initially its upward speed v is zero.

A hundredth of a second later its upward speed is −0.0981 m/s (i.e. downwards). The change dv = −0.0981 and the time interval dt = 0.01.

We can divide the change in speed by the change in time to get an acceleration −9.81 metres per second per second.

Now the initial height is z. School physics tells us that dt seconds later the height should be

$$z - 9.81 \ dt^2/2$$

But now the ratio of change-in-height to change-in-time for the initial speed will depend on dt

$$-9.81 \ dt/2$$

To get the answer 'zero' for the initial speed, we must look at the limit as dt gets very small, tending to zero. But for the purposes of engineering simulation, a value of 0.01 s will often give us an answer that is sufficiently accurate.

We have two differential equations

$$dz/dt = v$$

and

$$dv/dt = -9.81$$

In the computer, we can reverse the process of differentiation into one of integration. We can write two lines of code

```
z+ = v * dt;
v+ = -9.81 * dt;
```

where the first rather cryptic line means 'add v*dt to the value of z'. If we give dt the value 0.01 and repeat the calculation a hundred times per second, we will model the falling ball.

Look at www.essdyn.com/sim/falling.htm to see what I mean.

A.1.3 Partial Derivatives

We are all familiar with the potential energy of a ball being mgh. But as soon as we take the ball into space that all changes. We have axes x, y and z, but that does not necessarily tell us which direction is down!

It can be shown (it will become more obvious later) that the potential energy of something in the earth's gravitational field is

$$V = -mg \ R^2/r$$

relative to something that has 'escaped' to infinity, where R is the radius of the earth, r is the distance of the object from the centre of the earth and g is the familiar 9.81.

Now if we measure x, y and z from the centre of the earth, we have

$$r = \sqrt{(x^2 + y^2 + z^2)}$$

so for the force in the x direction we want to know the change in energy due to changing just x on its own

$$f_x = -\frac{\partial V}{\partial x}$$

$$= -\left\{ mgR^2 \frac{\partial}{\partial x} \left(-\frac{1}{r} \right) \right\}$$

$$= -mgR^2 \frac{x}{r^3}$$

If we want the whole force vector f, we can use the mathematicians' notation, using a bold x to represent the vector $(x, y, x)'$ and say that

$$\mathbf{f} = -\nabla V = -\frac{mgR^2}{r^3} \mathbf{x}$$

The reason we see the r cubed in the denominator instead of r squared is because we are using vector x/r to imbue the force with a direction. Of course, when we want to use it we will break it down into its separate equations.

To simulate an orbiting body, we have position coordinates held in variables x, y and z and velocities held in vx, vy and vz. At each time-step dt, we add dt times the appropriate acceleration to each velocity. We add $vx * dt$ to x, $vy * dt$ to y and $vz * dt$ to z. Then we move the picture of the object and do it all again.

For the acceleration, we can make mgR^2 a constant, maybe unity for simplicity, but we must calculate the components of $\mathbf{x}/r^3$.

To test it to see if it works look at www.essdyn.com/sim/orbitxyz.htm.

A.1.4   Another Example of Grad

Imagine a depth gauge on a rock under the sea. As the tide comes in and goes out, the reading will change. Its rate of change will be

$$\frac{\partial(depth)}{\partial t}$$

But now strap the gauge to the wrist of a diver who is swimming with vector velocity **v**. The depth will change at an additional rate determined by whether he is swimming upwards or down, i.e. by the scalar product of his velocity with the gradient of the depth function. Now depth is a simple scalar number, but the gradient of depth is a vector pointing straight down.

We end up with a total rate of change

$$\frac{\partial(\text{depth})}{\partial t} + \mathbf{v} \cdot \text{grad}(\text{depth})$$

or in a more mathematician-friendly form

$$\frac{d(\text{depth})}{dt} = \left(\frac{\partial}{\partial t} + \mathbf{v} \cdot \nabla\right)\text{depth}$$

We can express this more generally as an operator

$$\frac{d}{dt} = \frac{\partial}{\partial t} + \mathbf{v} \cdot \nabla$$

A.1.5 Swings and Roundabouts

For this, you really need to pay a visit to a children's playground. You should find a roundabout in the form of a flat wooden cylinder with handrails on top. As it spins, you will be readily aware of 'centrifugal force' (or 'centripetal acceleration'), but when you swing your leg in and out Coriolis force will make its presence felt as you kick your own ankle.

Suppose that your coordinates relative to the centre of the roundabout are given by a vector $\mathbf{r}$. Then your velocity $\mathbf{v}$ is given by $d\mathbf{r}/dt$. But when the roundabout spins, even though you are holding on to it without trying to move so that $\mathbf{r}$ remains constant, you have a velocity as your position vector sweeps round.

This velocity $\mathbf{v}$ will be the cross product of your relative position vector $\mathbf{r}$ with the roundabout's angular velocity vector

$$\mathbf{v} = \omega \times \mathbf{r}$$

But now suppose that you move about on the roundabout, you will add a further component

$$\frac{\partial \mathbf{r}}{\partial t}$$

so that your total velocity becomes

$$\frac{d\mathbf{r}}{dt} = \frac{\partial \mathbf{r}}{\partial t} + \omega \times \mathbf{r}$$

and we have a new operator

$$\frac{d}{dt} = \frac{\partial}{\partial t} + \omega \times$$

Now we can take an example from the mathematicians and use this operator on things other than the radius. In particular, we can differentiate your velocity to find your acceleration $d\mathbf{v}/dt$:

$$\frac{d\mathbf{v}}{dt} = \frac{\partial \mathbf{v}}{\partial t} + \omega \times \mathbf{v}$$

But we know that

$$\mathbf{v} = \left(\frac{\partial}{\partial t} + \omega \times \right) \mathbf{r}$$

so

$$\frac{d\mathbf{v}}{dt} = \frac{d}{dt} \frac{d\mathbf{r}}{dt}$$

thus the acceleration is

$$\frac{d^2\mathbf{r}}{dt^2} = \left(\frac{\partial}{\partial t} + \omega \times \right) \left(\frac{\partial}{\partial t} + \omega \times \right) \mathbf{r}$$

We must now pick this to pieces.

$$\frac{d^2\mathbf{r}}{dt^2} = \left(\frac{\partial}{\partial t} + \omega \times \right) \left(\frac{\partial \mathbf{r}}{\partial t} + \omega \times \mathbf{r} \right)$$

$$= \frac{\partial}{\partial t} \left(\frac{\partial \mathbf{r}}{\partial t} + \omega \times \mathbf{r} \right) + \omega \times \left(\frac{\partial \mathbf{r}}{\partial t} + \omega \times \mathbf{r} \right)$$

$$= \frac{\partial^2 \mathbf{r}}{\partial t^2} + \frac{\partial \omega}{\partial t} \times \mathbf{r} + \omega \times \frac{\partial \mathbf{r}}{\partial t} + \omega \times \left(\frac{\partial \mathbf{r}}{\partial t} + \omega \times \mathbf{r} \right)$$

$$= \frac{\partial^2 \mathbf{r}}{\partial t^2} + \frac{\partial \omega}{\partial t} \times \mathbf{r} + \omega \times \frac{\partial \mathbf{r}}{\partial t} + \omega \times \frac{\partial \mathbf{r}}{\partial t} + \omega \times (\omega \times \mathbf{r})$$

$$= \frac{\partial^2 \mathbf{r}}{\partial t^2} + \frac{\partial \omega}{\partial t} \times \mathbf{r} + 2\omega \times \frac{\partial \mathbf{r}}{\partial t} + \omega \times (\omega \times \mathbf{r})$$

So what do all these terms mean?

Well the first one is the acceleration with respect to the roundabout.

If the roundabout is accelerating, the second term tells us that our velocity will also accelerate.

But the third term depends on our velocity with respect to the roundabout. It is the **Coriolis force** that makes our leg swing to one side and kick our other ankle.

Now the final term might be a bit puzzling. It involves a 'vector triple product'. But we can take it gently.

The cross product of omega and **r** is a vector perpendicular to both of them, with size that is the product of their magnitudes times the sine of the angle between them.

In other words this is omega times the component of **r** resolved perpendicular to omega. It is the distance from the axis, rather than from the centre of the round-about coordinates.

When we take its cross product with omega again, its size will be multiplied by another omega and its direction will be towards the axis.

It is our good old friend 'centripetal acceleration' or 'centrifugal force'.

Appendix 2: Lagrange and Hamilton

Abstract In order to simulate a system we need some differential equations. Sometimes the system is simple and straightforward, so that the equations are clear to see, while sometimes they take more cunning. But be careful. Writers of erudite papers in learned journals prefer to write their equations in the most abstruse format they can find. The material that follows is more concerned with showing the relevance of advanced mathematics than with a simple 'how to do it'.

For example a simple pendulum swinging in a plane is most easily described via Newton's laws. If the mass is m, the tension in the string (for small swings) will be mg and when the pendulum swings to angle theta the restoring force is mg. sin(*theta*). Now if the linear deflection is x then (for small angles) *theta* $= x/L$, where the pendulum length is L. So, since the mass cancels out in the relationship between the force and the resulting acceleration, our equation is

$$\mathrm{d}^2x/\mathrm{d}t^2 = -g\,x/L$$

the classic equation of simple harmonic motion.

But below you will see some much more abstruse ways to get the equation involving energy or generalised coordinates.

A.2.1 Introduction

Whether solving a dynamic system analytically or preparing to simulate it, we need a mathematical representation of its variables. Usually the direct and intuitive way is by the application of Newton's laws. Sometime there are constraints or complexities that direct us to more general methods. But these methods lack the simplicity of Newton's laws. We can compare them with some simulation examples.

© Springer International Publishing AG 2018

J. Billingsley, *Essentials of Dynamics and Vibrations*,

DOI 10.1007/978-3-319-56517-0

A.2.2 Moon Lander Simulation

At www.essdyn.com/sim/lander.htm you will see a simple simulation of a moon-lander. The vehicle has been constrained to move in a plane so that it can be portrayed on a 2D screen. It involves the following variables:

Height
Vertical velocity
Horizontal position
Horizontal velocity
Tilt angle
Angular velocity
Fuel remaining.

Each one obeys a first order equation. Some of these are 'obvious', such as

$$d \, (\text{height})/dt = \text{vertical velocity}$$
$$d \, (\text{horizontal position})/dt = \text{horizontal velocity}$$
$$d \, (\text{tilt angle})/dt = \text{angular velocity}$$

For some of the others, we have Newton's laws that state that acceleration is proportional to force. There is also an angular version to tell us that angular acceleration is proportional to rotational torque. We can define the inputs, the thrust and the rotation torque, in terms that let us make the simulation constants unity:

$$d \, (\text{vertical velocity})/dt = \text{thrust} * \cos(\text{tilt}) - \text{lunar gravity}$$
$$d \, (\text{horizontal velocity})/dt = -\text{thrust} * \sin(\text{tilt})$$
$$d \, (\text{angular velocity})/dt = \text{torque}$$
$$d \, (\text{fuel})/dt = \text{thrust}$$

More often than not, a system will have nonlinear equations that cannot easily be solved without approximating them to something linear for small displacements. In the lander example, however, the horizontal acceleration is the thrust times the sine of the tilt angle.

Nonlinear equations present no problem at all when simulating the system. Time is advanced in steps of length dt. At each step the derivative*dt is added to each variable. By making dt sufficiently small, the solution can be made as accurate as needed.

But before we can start, we need to identify a set of 'state variables' like those above that describe exactly what is happening at any instant, plus a corresponding set of equations from which their rates of change can be calculated.

How can we find them?

A.2.3 Using Newton's Laws

These express the accelerations in terms of the applied forces.

With state variables in mind, we also consider the velocities so that the second-order equations are broken down into pairs of first order equations.

If a particle falls freely under gravity, we have

$$d^2y/dt^2 = -g$$

(if we plot y vertically in this 2D case) which we break down into

$$dy/dt = v$$

and

$$dv/dt = -g$$

To make this sort of analysis work, we need to use Cartesian coordinates and a set of axes that do not accelerate.

But many awkward problems involve rolling or spinning, where polar coordinates can make the analysis more straightforward.

A.2.4 System Energy and Virtual Work

If we have to represent the variables in terms of some other set of axes the analysis might be much less intuitive.

An approach that will work every time is to rely on the conservation of energy. An expression can be written for the total of kinetic and potential energy. Its rate of change will be the work done by any input forces. It sounds easy.

First look at it with simple axes.

A.2.4.1 Simplest Example

Let us consider that simple case of a mass m falling under gravity.

As before we consider vertical position and velocity y and v.

The kinetic energy K is $mv^2/2$. The potential energy U is mgy.

So the total energy, H, is given by

$$H = K + U$$
$$H = m v^2/2 + m g y$$

and with no other input to the system we have

$$dH/dt = 0$$
$$m \, v \, dv/dt + m \, g \, dy/dt = 0$$

But we know that dy/dt is the same as v, so

$$m \, v(dv/dt + g) = 0$$

thus

$$dv/dt = -g$$

This seems a very roundabout way to arrive at a result that was obvious from the start from Newton's laws. But notice that here we have not used Newton's laws at all – we have deduced the second law from conservation of energy.

For the kinetic energy, we could have thrown in the horizontal velocity as well, but we would have found that it was constant – or zero, as we have assumed here.

A.2.4.2   Second Example

Consider a simple pendulum. A point mass is suspended on a string of length L. How can we get an equation that is good for large displacements, rather than the simple equation

$$d^2x/dt^2 = -g \, x/L$$

that describes small ones?

We can take variables x, y, v_x and v_y, where x and y are measured relative to the centre of the pendulum. We can assume that the mass is unity.

The straightforward Newton approach is to consider the horizontal acceleration. We multiply the tension T in the string by the sine of the angle to resolve it to give the horizontal acceleration

$$T \sin(theta)/m.$$

But what is the tension? It is not a simple mg, or even necessarily something involving a sine or cosine. We will have to use some more ingenuity.

A.2.5   A Change of Variables

Although the pendulum bob has two coordinates x and y, these are coupled by the condition that

$$x^2 + y^2 + L^2$$

Instead we can use the single variable that is the angle theta of the pendulum, together with its rate of change omega.

Now if we think in terms of couples, the couple about the centre due to gravity is

$$-mgL \sin(\theta)$$

Since the string passes through that centre, the tension in it does not feature in the couple.

The moment of inertia of the pendulum about the centre is mL^2, so we end up with two equations

$$\frac{d\theta}{dt} = \omega$$

$$mL^2 \frac{d\omega}{dt} = -mgL \sin(\theta)$$

i.e.

$$\frac{d\omega}{dt} = -g \sin(\theta)/L$$

It took a bit of thought, so could there be some other general approach that can be used for every situation?

A.2.6   Considering Energy – The Hamiltonian

For the pendulum in a plane with y upwards, we have total energy

$$H = \frac{1}{2}mv_x^2 + \frac{1}{2}mv_y^2 + mgy$$

which we will call the Hamiltonian.

Now when we consider dH/dt, we must consider each contribution to its rate of change. We have

$$\frac{dH}{dt} = mv_x \frac{dv_x}{dt} + mv_y \frac{dv_y}{dt} + mv_y g = 0$$

which looks exactly the same as the case for the freely falling mass.

But we also have the condition

$$x^2 + y^2 = L^2$$

together with the fact that the horizontal acceleration is no longer zero. How can we combine the two conditions?

Differentiating this second condition gives

$$2xv_x + 2yv_y = 0$$

which we can differentiate a second time to get

$$2v_x^2 + 2v_y^2 + 2x\frac{dv_x}{dt} + 2y\frac{dv_y}{dt} = 0$$

It is looking hopeful, but will clearly need a heap of algebra to sort it out. Maybe we need to look at the polar coordinates again.

A.2.7   Generalised Coordinates

Instead of looking at everything that can move, we can just consider variables that allow the 'degrees of freedom' of our system.

In the case of the pendulum, this is just the angle of swing, together with any necessary rates of change.

The total energy is once again

$$H = \frac{1}{2}mv^2 + mgy$$

where v is the magnitude of the combined velocity, but now

$$v = L\omega$$

where

$$\omega = \frac{d\theta}{dt}$$

and

$$y = L(1 - \cos(\theta))$$

so when we use

$$dH/dt = 0$$

we get as before

$$mv\frac{dv}{dt} + mg\frac{dy}{dt} = 0$$

which when we substitute $L\omega$ for v now gives

$$mL\omega L\frac{d\omega}{dt} + mgL\sin(\theta)\omega = 0$$

so

$$mL\omega\left(L\frac{d\omega}{dt} + g\sin(\theta)\right) = 0$$

and once again

$$\frac{d\omega}{dt} = -\frac{g}{L}\sin(\theta)$$

i.e.

$$\frac{d^2\theta}{dt^2} = -\frac{g}{L}\sin(\theta)$$

It still seems a lot of trouble to go to get a simple result. But how about looking at a problem where the result is not so simple!

A.2.8 Ball on an Elastic String

The kinetic energy will still be

$$\frac{1}{2}mv^2$$

but now

$$v^2 = (L\omega)^2 + \left(\frac{dL}{dt}\right)^2$$

where

$$\omega = \frac{d\theta}{dt}$$

and the potential energy has to include the 'stretch' of the elastic.

Let us specify that when the ball is at rest, the elastic is stretched by the weight of the ball to twice its unstretched length L_0.

It is not hard to show that this part of the potential energy is

$$\frac{mg(L-L_0)^2}{2L_0}$$

while the 'height' component measured from the suspension point is

$$-mgL\cos(\theta)$$

For the Hamiltonian H we get

$$m\left\{\frac{1}{2}L^2\omega^2 + \frac{1}{2}\left(\frac{dL}{dt}\right)^2 + \frac{1}{2}g\frac{(L-L_0)^2}{L_0} - Lg\cos(\theta)\right\}$$

Now we have to start differentiating. But remember that L is now a variable, not a constant. Since dH/dt is zero, we can divide through by m. Taking the derivatives of each of the four terms, we have

$$L\frac{dL}{dt}\omega^2 + L^2\omega\frac{d\omega}{dt}$$

$$+ \frac{d^2L}{dt^2}\frac{dL}{dt}$$

$$+ g\frac{L-L_0}{L_0}\frac{dL}{dt}$$

$$+ Lg\sin(\theta)\frac{d\theta}{dt} - g\cos(\theta)\frac{dL}{dt}$$

and the whole total must be zero.

But this gives us just one equation for the second derivatives and we need two.

A.2.9   The Lagrangian and 'Action'

It is intuitive to add the potential energy to the kinetic energy and use conservation of energy. But Lagrange took their difference and called it 'action'. He asserted that a dynamic system would behave in a way that minimised the integral of the action.

The Lagrangian is defined as

$$L = K - U$$

where K is the kinetic energy and U is the potential energy. He expressed these in terms of a set of 'generalised coordinates' q_1, q_2 and so on.

Now this sort of minimisation requires the use of 'Calculus of Variations' for its proof. The net result is as follows:

$$\frac{\mathrm{d}}{\mathrm{d}t}\left[\frac{\partial L}{\partial \dot{q_i}}\right] - \frac{\partial L}{\partial q_i} = 0$$

Will it succeed in cracking the 'ball on elastic' example?

We now have length L and angle theta in place of q_1 and q_2. But to use L for the length will be confusing, since we have now used L for the Lagrangian!

So let us instead represent the length as r.

L is given by $K - U$ which is here

$$m\left\{\frac{1}{2}r^2\omega^2 + \frac{1}{2}\left(\frac{\mathrm{d}r}{\mathrm{d}t}\right)^2 - \frac{1}{2}g\frac{(r-r_0)^2}{r_0} + rg\cos(\theta)\right\}$$

so by applying Lagrange's equations for both r and theta we will get the two second order differential equations that we need. But instead we will represent them as four first-order equations.

The first two equations are obvious.

If we write dr/dt as v (take care, it is not the total velocity) then we have

$$\mathrm{d}r/\mathrm{d}t = v$$

and

$$\mathrm{d}\theta/\mathrm{d}t = \omega$$

Our partial differential equations become

$$\frac{\mathrm{d}}{\mathrm{d}t}\left[\frac{\partial L}{\partial v}\right] - \frac{\partial L}{\partial r} = 0$$

and

$$\frac{\mathrm{d}}{\mathrm{d}t}\left[\frac{\partial L}{\partial \omega}\right] - \frac{\partial L}{\partial \theta} = 0$$

where the Lagrangian L is now

$$L = m \left\{ \frac{1}{2} r^2 \omega^2 + \frac{1}{2} v^2 - \frac{1}{2} g \frac{(r - r_0)^2}{r_0} + rg \cos(\theta) \right\}$$

Taking the equations step by step we see that

$$\frac{\partial L}{\partial v} = mv$$

and

$$\frac{\partial L}{\partial r} = m \left\{ r \omega^2 - g \frac{(r - r_0)}{r_0} + g \cos(\theta) \right\}$$

to help with the first equation and

$$\frac{\partial L}{\partial \omega} = mr^2 \omega$$

with

$$\frac{\partial L}{\partial \theta} = m \{ -rg \sin(\theta) \}$$

for the second.

Putting these together, after applying the necessary 'd/dt's and dividing through by m we get

$$\dot{v} = r\omega^2 - g \frac{(r - r_0)}{r_0} + g \cos(\theta)$$

which will be our third state equation, while

$$r^2 \dot{\omega} - 2rv\omega = -rg \sin(\theta)$$

can be tidied up to give the fourth as

$$\dot{\omega} = -2v\omega / r - g \sin(\theta) / r$$

Now we are all set to write a simulation, or even perhaps to solve the equations analytically.

But we have to ask whether the Lagrangian gives an easy answer. You will find it widely used in journal papers where the author is perhaps trying to impress the reader.

Let us compare this with an alternative approach.

A.2.10   Direct Cartesian Calculation

For a straightforward simulation we would like to use variables x and y, together with their derivatives v_x and v_y.

The first two equations are simply

$$dx/dt = v_x$$

$$dy/dt = v_y$$

But for the others we need to calculate the accelerations. In the x direction the acceleration is caused by the tension in the elastic. In the y direction we have the tension and gravity. We can save some typing by taking the mass of the ball as unity.

$$dv_x/dt = -\text{tension} \sin(\text{theta})$$

$$dv_y/dt = \text{tension} \cos(\text{theta}) - g$$

But first we have to work out the tension and theta. We start with L, the length of the elastic. If we measure this from the suspension point, we simply have

$$L^2 = x^2 + y^2$$

Now for the tension, we can save even more typing by taking $L_0 = 1$.
For a stretch of 1 the tension is g, so the tension for length L is

$$\text{tension} = g(L - 1)$$

But in our simulation we can do better than that! We can say that if $L < 1$ the elastic has gone slack and the tension is zero.

And now that we know L, we also know that

$$\sin(\text{theta}) = x/L$$

$$\cos(\text{theta}) = -y/L$$

(since we are measuring y upwards and the ball will hang down)

Putting these all together we have some code to use in a simulation

```
length = Math.sqrt(x*x+y*y);
if(length>1){
    tension = g*(length - 1);
}else{
    tension = 0;
}
vx -= dt * tension * x/length;
vy -= dt * (tension * y/length + g);
x += dt * vx;
y += dt * vy;
MoveBall(x,y);
```

You can see it in action at www.essdyn.com/sim/elastic.htm.

The simple method can often be the best!

Indeed it is easy to see how to extend the simulation to the three dimensional one at www.essdyn.com/sim/elastic3D.htm.

Further Reading

Here are some titles for further reading that I have found, though I do not express a preference for any particular one of them.

Hibbeler RC, Yap KB (2013) Mechanics for Engineers: DYNAMICS, 13th or 14th edn. Pearson, London, UK

Hibbeler RC (2017), Engineering Mechanics: Dynamics, 13th edn. Pearson, London, UK, p. 768

Meriam JL, Kraige LG (2007) Engineering mechanics, vol II. Dynamics, 6th edn. Wiley, New Jersey, USA

Beer FP, Johnston ER (2005) Vector mechanics for engineers: Dynamics, 6th edn. McGraw-Hill, New York, USA

Ginsberg J (2008) Engineering Dynamics, 3rd edn. Cambridge University Press, Cambridge, UK

Greenwood DT (2003) Advanced Dynamics, Cambridge University Press, Cambridge, UK

Kasdin NJ, Paley DA (2011) Engineering Dynamics: A Comprehensive Introduction, Princeton University Press, Princeton, USA

O'Reilly OM (2010) Engineering Dynamics: A Primer, 2nd ed. Springer-Verlag, New York, USA

Synge JL, Griffith BA (1949) Principles of Mechanics, McGraw-Hill, New York, USA, can be downloaded from https://archive.org/details/principlesofmech031468mbp

© Springer International Publishing AG 2018
J. Billingsley, *Essentials of Dynamics and Vibrations*,
DOI 10.1007/978-3-319-56517-0

Further Reading

Printed in the United States
By Bookmasters